EVOLUTIONARY DIFFERENTIATION IN MORPHOLOGY, VOCALIZATIONS, AND ALLOZYMES AMONG NOMADIC SIBLING SPECIES IN THE NORTH AMERICAN RED CROSSBILL (*LOXIA CURVIROSTRA*) COMPLEX

Evolutionary Differentiation in Morphology, Vocalizations, and Allozymes Among Nomadic Sibling Species in the North American Red Crossbill (*Loxia curvirostra*) Complex

Jeffrey G. Groth

A Contribution from the Museum of Vertebrate Zoology
of the University of California at Berkeley

UNIVERSITY OF CALIFORNIA PRESS
Berkeley • Los Angeles • London

UNIVERSITY OF CALIFORNIA PUBLICATIONS IN ZOOLOGY

Editorial Board: Peter B. Moyle, James L. Patton,
Donald C. Potts, David S. Woodruff

Volume 127
Issue Date: June 1993

UNIVERSITY OF CALIFORNIA PRESS
BERKELEY AND LOS ANGELES, CALIFORNIA

UNIVERSITY OF CALIFORNIA PRESS, LTD.
LONDON, ENGLAND

© 1993 BY THE REGENTS OF THE UNIVERSITY OF CALIFORNIA
PRINTED IN THE UNITED STATES OF AMERICA

Library of Congress Cataloging-in-Publication Data

Groth, Jeffrey G.
    Evolutionary differentiation in morphology, vocalizations, and allozymes among nomadic sibling species in the North American red crossbill (Loxia curvirostra) complex / Jeffrey G. Groth.
    p.   cm. — (University of California publications in Zoology; v. 127)
    "A contribution from the Museum of Vertebrate Zoology of the University of California at Berkeley."
    Includes bibliographical references.
    ISBN 0-520-09782-3 (pbk. : alk. paper)
    1. Red crossbill—North America—Morphology. 2. Red Crossbill—North America—Vocalization. 3. Red crossbill—North America—Classification. I. University of California, Berkeley. Museum of Vertebrate Zoology. II. Title. III. Series.
QL696.P246G76 1993
598.8'83—dc20   93-17021
                                                                       CIP

The paper used in this publication meets the minimum requirements of American National Standard for Information Sciences—Permanence of Paper for Printed Library Materials, ANSI Z39.48-1984. ♾

Contents

List of Figures, vii
List of Tables, ix
Acknowledgments, xi

INTRODUCTION ... 1
    Synopsis of Taxonomy and Natural History, 2
    Summary, 7

MATERIALS AND METHODS ... 8
    Specimens Examined and Study Sites, 8
    Vocalizations, 11
    Morphology, 13
    Plumage Color Description, 16
    Allozyme Electrophoresis, 17

COMPARISON OF MUSEUM SPECIMENS OF CROSSBILLS
TO SPECIMENS OF KNOWN VOCALIZATIONS ... 20

THE VOCAL REPERTOIRE ... 26
    Calls of Adults, 26
    Calls of Juveniles, 32

GROUPS BASED ON VOCALIZATIONS ... 36
    Descriptions of Flight Call Types, 36
    Correspondence Between Flight Calls and Other Calls, 39
    Flight Call Variation Within Geographic Regions, 41
    Summary, 51

SOURCES OF INTRAPOPULATION VARIATION IN MORPHOLOGY 55
Variability of Morphological Characters, 55
Morphological Geographic Variation in a Single Vocal Group, 60
Sexual Dimorphism, 60
Variation in Size with Age, 62
Discussion, 67

MORPHOLOGICAL DIFFERENCES AMONG VOCAL GROUPS 68
Univariate Patterns, 68
Principal Components Analysis, 71
Morphological Discrimination of Vocal Groups, 75
Summary, 78

COLOR VARIATION 79

ECOLOGICAL VARIATION 82

ALLOZYME VARIATION 84
General Patterns of Allozyme Variation, 84
Comparison of *L. curvirosta* and *L. leucoptera*, 85
Allozyme Comparisons of *L. curvirostra* Vocal Groups, 86
Discussion, 88

DISCUSSION AND CONCLUSIONS 92
Level of Differentiation, 92
Speciation in North American Red Crossbills, 93
Nomenclature, 95
Summary, 97

Appendix A: List of Field Localities and Sample Sizes, 99
Appendix B: Collections and Sample Sizes of Museum Specimens of *L. curvirostra* Examined in This Study, 102
Appendix C: List of Recordings Made by Other Observers, 103
Appendix D: Univariate Measurements of Adults in 7 Vocal Groups of Red Crossbills, 105
Appendix E: Individual Identification Numbers, Catalog Numbers, and Vocal Group Numbers for the Sample of Red Crossbills with Recorded Vocalizations, 111
Appendix F: Allozyme Frequencies for Polymorphic Loci in *L. leucoptera* and 7 Samples of *L. curvirostra*, 116

Literature Cited, 121

List of Figures

1. History of nomenclature in North American *Loxia curvirostra*, 4
2. Map of recording regions and localities, 9
3. Map of museum specimen coverage in this study, 10
4. Method of bill curvature measurement, 14
5. Morphological comparison of museum specimens with crossbills of known vocalizations, 22
6. Clinal variation in morphology of museum specimens of crossbills from western North America, 23
7. Description of the flight call, 27
8. Alarm, excitement, and distress calls, 29
9. Miscellaneous vocalizations, 31
10. Ontogeny of calls in nestlings, 33
11. Call learning in juveniles, 34
12. Division of flight calls into groups, 37
13. Division of alarm and excitement calls into groups, 40
14. Flight calls of crossbills in the northwest region, 42
15. Flight calls of crossbills in Oregon, 44
16. Flight calls of crossbills in northern and coastal California, 45
17. Flight calls of crossbills in the Sierra Nevada, California, 46
18. Flight calls of crossbills in Arizona, 47
19. Flight calls of crossbills in Colorado and New Mexico, 49
20. Flight calls of crossbills in the southern Appalachian region, 50
21. Flight calls of crossbills in the northeast region, 52
22. Geographic distributions of eight vocal groups of crossbills, 53
23. Bill variation with wear, 58
24. Geographic variation in a single form of red crossbill, 63
25. Correlation between skull pneumatization and plumage development, 65
26. Development of size in juveniles, 66
27. Photograph of study skins of seven different red crossbill forms, 69

List of Figures

28. Wing size and shape, 70
29. Bill curvature distributions, 71
30. Principal components plots for bill and skeletal measurements, 74
31. Canonical discriminant analysis plots of study skin measurements, 77
32. Distribution of color variants, 80
33. Phenograms portraying allozyme distances, 89
34. Correlation between genetic and morphological distances, 90

List of Tables

1. Plumage scoring scales for age estimation of juveniles and color description of adult males, 11
2. Correlation coefficients among museum study skin characters, 21
3. Principal components analysis of museum specimens, 21
4. Clinal variation in museum specimens of North American red crossbills, 24
5. Coefficients of variation (CVs) and sexual dimorphism, 56
6. Leg asymmetry in crossbills, 59
7. Correlation coefficients among morphological characters within a single form of red crossbill, 61
8. Principal components analysis within a single form of red crossbill, 62
9. Principal components analysis using bill characters, 72
10. Principal components analysis using skeletal characters, 73
11. Discriminant function coefficients separating 7 forms of red crossbill, 76
12. Mahalanobis' distances based on morphological characters, 78
13. Genetic parameters of *L. curvirostra* and *L. leucoptera*, 87
14. Genetic distances among samples of crossbills, 88

Acknowledgments

Financial support was provided by the Frank M. Chapman Memorial Fund of the American Museum of Natural History, the Alexander Wetmore Fund of the American Ornithologists' Union, the Kellogg Fund of the Museum of Vertebrate Zoology, the Department of Zoology of the University of California at Berkeley, Sigma Xi, and an NSF Dissertation Improvement Grant (BSR-8700999).

I am especially indebted to Curtis S. Adkisson for sharing his expertise and for his loan of captive crossbills. I thank Ned K. Johnson for directing the final years of the project. James L. Patton and Thomas Duncan provided valuable counsel whenever needed. Robert E. Jones expertly prepared all of the skeletal specimens. Jill Marten and Monica Frelow helped with the electrophoretic work. Effie Dilworth, Pat Kelly, and Chris Meacham assisted with various aspects of computerization of data. Shelley Kahane and Robert Baiz helped care for the captive birds. Curtis S. Adkisson, Douglas McNair, Daniel Rosenburg, Kathleen Groschupf, and John Trochet led me to crossbills at various field sites. Walter Bock sent a number of frozen white-winged crossbills which were used in the allozyme survey. Carla Cicero helped in packing and unpacking loans of museum specimens. I thank my parents for their moral support over the years, and I give special thanks to Sharon L. Stein for helping in many ways.

The critical comments contributed by Ned K. Johnson, James L. Patton, Robert M. Zink, Peter B. Moyle of the editorial board, and an anonymous reviewer, substantially improved the quality of the analysis and text. The assistance of Rose Anne White of the University of California Press was essential in compiling the final version of this monograph.

The following people generously made tape recordings available from personal collections or from archives under their administration: Curtis S. Adkisson, Craig Benkman, James L. Gulledge (with Kate O'Hara at the Cornell Laboratory of Ornithology), Tom Hahn, J. W. Hardy (Florida State Museum collection), Ned K. Johnson, Donna J. Lusthoff, Marie L. Mans, Joe T. Marshall, Charles Nicholson, Robert B. Payne (with Robert Preston), Elinor Pugh, Robert N. Randall, George B. Reynard, William S. Shepherd, and Cynthia A. Staicer.

The Museum of Vertebrate Zoology and Department of Zoology at the University of California provided excellent laboratories, offices, and staff. Facilities for keeping captive birds were provided by the Field Station for Behavioral Research of the University of California at Berkeley. Highlands Biological Station and Sagehen Creek Field Station provided housing for some of the field work. I am also grateful for the support from the American Museum of Natural History (Frank M. Chapman Fund) during the phases of revision and production of the final copy.

The courtesy of the staff of the following research collections made it possible to examine specimens either through loans or visits: American Museum of Natural History (Robert W. Dickerman), Burke Museum at the University of Washington (Sievert A. Rohwer, Carol Spaw and Chris Wood), California Academy of Sciences (Luis Baptista and Steve Bailey), California State University at Long Beach (S. Warter), Carnegie Museum of Natural History (Kenneth C. Parkes), Cornell University (Kevin J. McGowan), Delaware Museum of Natural History (David M. Niles and Gene K. Hess), Denver Museum of Natural History (Joanne Carter), Museum of Vertebrate Zoology (Ned K. Johnson), University of Kansas Museum of Natural History (Tristan J. Davis), Louisiana State University Museum of Natural History (J. Van Remsen), National Museum of Canada (Henri Ouellet), Royal British Columbia Museum (R. Wayne Campbell and Grant W. Hughes), San Diego Natural History Museum (Amadeo Rea and Stephen Gustafson), Santa Barbara Museum of Natural History (Paul W. Collins and Kathryn A. Barry), University of Arizona (Tom Huels), University of California at Los Angeles (James R. Northern), University of California at Santa Barbara (Mark Holmgren), University of Montana (T. J. Pratt and DeWayne Williams), University of Nebraska State Museum (Thomas E. Labedz), University of Puget Sound (Susan F. McMahon), University of Wisconsin at Madison (Frank I. Iwen), and Washington State University (Richard E. Johnson).

Scientific collecting permits were granted by the U. S. Fish and Wildlife Service and the fish and game agencies of the following states and provinces: British Columbia, Washington, Oregon, California, Idaho, Nevada, Arizona, Montana, Colorado, New Mexico, Ontario, Minnesota, Michigan, Quebec, New York, Virginia, North Carolina, New Hampshire, the Atlantic Provinces region, and Newfoundland. Specimens have been deposited in the Museum of Vertebrate Zoology, the American Museum of Natural History, and the Bailey-Law Collection at Virginia Polytechnic Institute.

INTRODUCTION

Birds are among the most intensively studied groups of organisms, yet the status, species limits, and geographic distributions of many forms remain poorly understood. Although this lack of knowledge is most serious for tropical avifaunas, a number of groups from temperate regions present unsolved problems. One of the most difficult and perplexing issues in the systematics of Holarctic birds concerns the morphologically variable red crossbill complex (*Loxia curvirostra*, Aves: Carduelinae). It is currently controversial whether this complex represents a single geographically variable species (Griscom 1937, American Ornithologists' Union 1957, Mayr and Short 1970, Monson and Phillips 1981, Dickerman 1986a); a series of sibling species (Groth 1988); or a more complicated combination of these alternatives.

Crossbills are nomadic and wander over broad continental areas; thus explanations accounting for their morphological diversification are difficult. Previous studies of crossbill systematics used museum specimens, but little agreement has been reached on the status, distribution, and mode of evolution in this complex. A new approach using vocalizations combined with morphology at the level of individual birds showed two forms of red crossbills breeding in the southern Appalachians (Groth 1988). Although morphologically distinctive, the two forms spanned only part of the range of bill and body size in the complex, therefore it became clear that a study combining morphological and vocal data on crossbills in other regions of North America was needed.

The purposes of this study are: (1) to summarize the natural history and taxonomy of red crossbills and outline outstanding systematic problems; (2) to provide a quantitative overview of patterns of morphological variation in the complex; (3) to document major features of the vocal repertoire, and identify those vocalizations that might be useful in comparing different populations; (4) to determine whether either of the two Appalachian forms exist in other regions and examine the extent to which forms of crossbill, identified on the basis of vocalizations, intergrade in morphology, color, and allozymes; (5) to suggest hypotheses regarding the evolution and status of crossbill forms; and (6) to show how new information derived from crossbills of known vocalizations and morphology might be used to solve current taxonomic problems in the complex.

SYNOPSIS OF TAXONOMY AND NATURAL HISTORY

General natural history. Red crossbills are widespread throughout the Holarctic where they specialize on conifer seed extraction with their crossed bills. Bill size in populations corresponds to the size and robustness of conifer cone types (Kirikov 1940; Lack 1944a, 1944b; Phillips et al. 1964; Newton 1973). The smallest forms occur in disjunct populations in the Himalayas (*Loxia curvirostra himalayensis*), the Philippines (*L. c. luzonensis*), and North America (*L. c. minor*). The largest forms are found in Europe (*L. c. curvirostra*), Vietnam (*L. c. meridionalis*), and southern North America (*L. c. stricklandi*). Specimens from Europe and North America show a wide range of bill and body size both within regions and within many series (Griscom 1937, Monson and Phillips 1981, Payne 1987).

Many ornithologists have described the erratic and unpredictable occurrence of red crossbills (e.g. Griscom 1937; Bent 1968; Newton 1970, 1973; Monson and Phillips 1981; Payne 1987). Furthermore, unlike virtually all north temperate birds, *L. curvirostra* nests in all seasons, including winter at high latitudes and altitudes (see McCabe and McCabe 1933, Bailey et al. 1953). Breeding and movements appear to be mainly in response to conifer cone crops (Reinikainen 1937, Lawrence 1949, Bailey et al. 1953, Newton 1970, Bock and Lepthien 1976, Benkman 1990), although day length affects testis size in captive males (Tordoff and Dawson 1965); therefore, the proximate factors inducing breeding are not fully understood. Crossbills may be entirely absent from broad regions over several years if cone crops are not available, and then "invade" in large numbers when new conifer seeds develop.

The Old World red crossbill complex. The nearest relatives of *L. curvirostra* are two large-billed Old World forms: the parrot crossbill (*L. pytyopsittacus*) and the Scottish crossbill (*L. scotica*). These closely resemble *curvirostra* in plumage. Early descriptions (Brehm 1853) showed that continental European crossbills formed a continuously graded series in bill size. Brehm recognized 11 different forms in the *curvirostra–pytyopsittacus* complex, and he described size, color, and some vocal differences among them. Hartert (1904) viewed *curvirostra* and *pytyopsittacus* as separate species, each divided into subspecies. After examination of over 500 Eurasian and North African specimens, most from the Rothschild Collection at the American Museum of Natural History in New York, Griscom (1937) considered the entire complex to be a single species, *L. curvirostra*, even though he was aware of the broad sympatry and ecological segregation of large- and small-billed forms in northern Europe. Griscom wrote that measurements taken on specimens from Cyprus (*L. c. guillemardi*) were almost identical to those of specimens from Scotland (*L. c. scotica*), and that these races formed morphological intermediates between nominate *curvirostra* and *pytyopsittacus*. The view that the complex was a single species was accepted by Meinertzhagen and Williamson (1953), but not by Witherby et al. (1938), the British Ornithologists' Union (1956), Vaurie (1959), nor Voous (1960). The taxonomic position of large-billed *scotica* has been either as a form of *curvirostra* (e.g., Hartert 1904, British Ornithologists' Union 1934, Griscom 1937, Witherby et al. 1938, Vaurie 1959, Voous 1960) or as a form of *pytyopsittacus* (e.g., Hartert and Steinbacher 1932,

British Ornithologists' Union 1956). Knox (1975) and Voous (1978) have argued that *scotica* has not hybridized significantly with *curvirostra* where the two are sympatric, and that the two forms have maintained their morphological distinctiveness even with ample opportunity for interbreeding. Further data showed that the two forms use the same forests for breeding (Knox 1990). Nethersole-Thompson (1975) documented vocal and ecological differences between *scotica* and *curvirostra* in his extensive biological investigations of crossbills in Scotland, and believed that *scotica* was closer to *pytyopsittacus* than to *curvirostra*.

New World red crossbills. Concepts on the taxonomic division of North American *L. curvirostra* have fluctuated widely over the last century (Fig. 1). The complex was first recognized as two species: a small northeastern *Curvirostra americana* Wilson 1811 and a larger southern *Loxia mexicana* Strickland 1851. Coues (1874) and Baird et al. (1874) considered *americana* and *mexicana* as races of *L. curvirostra,* uniting North American red crossbills with Old World populations. Ridgway (1885a) had "long suspected the existence of two forms of the red crossbill in the United States, besides the Mexican race" (p. 101), and he gave measurements showing that specimens collected at Fort Klamath, Oregon, were intermediate in size between *mexicana* and *americana*. The Oregon form was named *L. c. bendirei* Ridgway 1885. Citing Brehm's descriptions of *Crucirostra minor*, Ridgway (1885b) replaced the northeastern *americana* with *L. c. minor*. In the same publication, Ridgway stated that *L. mexicana* was preoccupied (presumably by *L. mexicana* Linnaeus 1758), and he therefore substituted the name *L. c. stricklandi* for the Mexican form.

New subspecies were added to the North American complex beginning with Grinnell's (1909) description of *L. c. sitkensis*, a small form with orange-colored males from Alaska. This was followed closely by Bent's (1912) description of *L. c. percna,* a large form on Newfoundland.

A few years later, Stresemann (1922) examined four specimens at the Berlin Museum that were presumably used by Brehm in his description of *minor*. Stresemann noted that Gloger (1834) had applied two names, "*Loxia pusilla* Lcht." and "*Crucirostra americana* Wls.," to the same group of specimens. Gloger did not specify which specimens belonged to which names, and Stresemann restricted the name *pusilla* to the larger male (number 6984), which he designated as a lectotype. The AOU (1931) resurrected the name *L. c. pusilla* Gloger 1834 for the small northeastern subspecies because of seniority in publication over Brehm 1846 (although Brehm 1845 is actually the earliest reference to the name *minor,* see Payne 1987). Van Rossem (1934) later visited the Berlin Museum, measured the four specimens examined by Stresemann, and concluded that two distinct size classes of birds were represented. This was apparently in agreement with Gloger's (1834) application of two names for the birds. Van Rossem concluded that Brehm's (1853) figure of crossbill head and bill profiles which included "den Typus von *L. c. minor*" was of a very small individual, although Brehm did not indicate which specimen was the type. Because the larger specimens were now *pusilla* according to Stresemann (1922), van Rossem (1934) designated number 6982 as the restricted type (lectotype) for *minor*. Although van Rossem did not directly compare the Berlin specimens to newer North

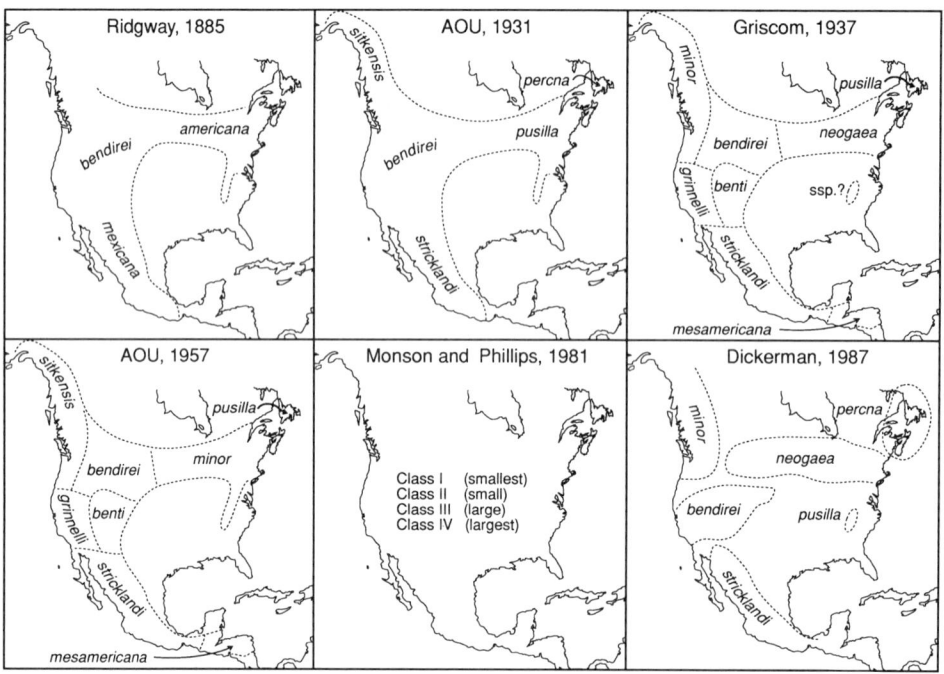

Figure 1. Concepts of subspecific division of the North American *Loxia curvirostra* complex over the last century.

American birds, he regarded *pusilla* as morphologically matching the large-billed birds of Newfoundland. This led him to replace *percna* with *pusilla* through priority, even though Stresemann's (1922) inquiry led to the conclusion that the type of *pusilla* came from Georgia, not Newfoundland.

Griscom's (1937) major revision of crossbill taxonomy further divided the complex into more subspecies. The only consistency between Griscom and the AOU (1931) was that *L. c. stricklandi* was maintained as the name for the large Mexican form. Western crossbills under the name *bendirei* were split into three subspecies: *L. c. bendirei* Ridgway; *L. c. grinnelli* subsp. nov.; and *L. c. benti* subsp. nov. The differences among these were based primarily on color and slight differences in size and shape. "Dark" birds smaller than *stricklandi* from countries south of Mexico were designated *L. c. mesamericana* subsp. nov. Griscom agreed with van Rossem (1934) that *pusilla* should be the name for Newfoundland birds (although Griscom never directly examined the type of *pusilla*). This decision was based on his belief that most large-billed specimens collected in eastern North America were migrants from the north (Newfoundland), and therefore it was not unlikely that the type came from Georgia. Continuing controversy surrounds Griscom's synonymy of *minor* and *sitkensis* (with *minor* taking priority). The rationale behind this decision involved the speculation that the very smallest New

World red crossbills (*sitkensis* Grinnell 1909) originate in the Pacific Northwest and invade eastward. Using the photograph of the type specimens of *minor* and *pusilla* published by van Rossem (1934) as a reference, Griscom noted that the type of *minor* was smaller than most specimens from eastern North America and was therefore a migrant from the far west. However, neither Griscom nor van Rossem knew the type locality of the specimen because the tag contained only "Nord-America." Griscom's interpretation meant that both type specimens previously used to represent eastern North American forms (*minor* and *pusilla*) were taken thousands of kilometers from their normal ranges. Each represented a race which was not the common eastern crossbill. This led Griscom to create a new name, *L. c. neogaea* subsp. nov., for a medium-sized crossbill in the upper Canadian zone of eastern North America and the Great Lakes region. Crossbills collected in eastern Tennessee were hypothesized to belong to a subspecies separate from the adjacent northern *neogaea* because "adult males [are] chiefly scarlet instead of brick red; juvenal darker and blacker than any other subspecies" (p. 115), but Griscom did not propose a new name for this suspected form.

Griscom discussed at length the great amount of morphological variation within North American regions. For eastern North America he wrote: "It is perfectly well known that there are three 'types' of red crossbills in the enormous series from southern New England and the Atlantic states" (p. 92). The smallest and largest forms were thought to be vagrants from the Pacific Northwest and Newfoundland, respectively, with only the medium-sized birds representing the regional subspecies. According to Griscom, breeding of one subspecies within the range of another often occurred, but races returned to their own areas after such incidents.

The AOU (1957) followed Griscom's revision with one exception. The name *neogaea* was not accepted, and *minor* and *sitkensis* were not synonymized. The name *minor* was reinstated for an eastern North American form, and *sitkensis* was used for a Pacific coastal race.

The most extensive museum work on North American *L. curvirostra* was done by Allan R. Phillips during the 1970s. A great many study skin specimens in major collections in the United States received penciled "ARP" references and subspecies identifications. Phillips (1975:282) wrote: "though all races are similar in voice and biology (breeding, plumages) so far as known, they may have widely overlapping regular ranges with apparently little cross-breeding," and further: "they may nest at any season, but chiefly in winter and sometimes on the other side of the continent from their usual home range." A non-binomial classification system of four "size classes" (I, II, III, and IV, from smallest to largest) was published as an appendix in Monson and Phillips' (1981) second edition of *Checklist of the Birds of Arizona*. They stated that Arizona has "independent little semi-species wandering around the state," and "intermediates are relatively infrequent" (p. 223). The four size classes were further divided on the basis of color and slight differences in bill shape. One of these variants, a more richly-colored class II (labeled "Ph") from Colorado, was given the name *L. c. vividior* subsp. nov. Another name, *L. c. reai* subsp. nov., was given to a dull-colored class I form, the type series having been taken in Idaho. There was some agreement with Griscom's (1937) synonymy of *minor* and *sitkensis* (they were both considered "Size Class I"), but not with the use of the name *pusilla* (which was placed within "Size Class II") for Newfoundland birds (*percna* was placed in "Size Class III"). This taxonomic system allowed geographic overlap of distinctive forms of crossbills, although the complex was not divided into separate species. However, Monson and Phillips suggested that different forms of crossbill may be reproductively isolated, but that "similarity in biology as a whole" among the forms was too great to warrant division into different species.

Payne (1987) compared the type specimens of *minor* and *pusilla* directly to specimens from North America. Measurements of *pusilla* fell outside the range of variation of Newfoundland crossbills, but they were similar to measurements taken on specimens from the type localities of the races *benti* (Grafton, North Dakota), *bendirei* (Fort Klamath, Oregon), and *neogaea* (Lake Umbagog, Maine), suggesting synonymy of all four. The names *minor, sitkensis,* and *reai* were found to represent morphologically indistinguishable forms. Payne determined that the type locality and date of collection for the type of *minor* was the Black River of Michigan in 1834, and that available specimens provided evidence for regular breeding of very small crossbills in Michigan. The total sample from Michigan was extremely variable in morphology, spanning the entire range of variation among all specimens from North America used in his study.

Dickerman (1986a, 1986b, 1987) recently proposed another system of names for North American red crossbills. Lines on his map indicate distributions of "core ranges" in which distinctive morphological forms usually breed. He described how subspecies occasionally leave their "core ranges" in times of cone crop failure and then breed within the ranges of other subspecies. Dickerman (1987) compared numbers of "*neogaea*" versus "non-*neogaea*" specimens collected in New England over the last century. The results suggested that various morphotypes of red crossbills have responded numerically in different ways to logging practices over the years, with the near-extinction of an "Old Northeastern" (medium-sized) subspecies around the turn of the century.

My study (Groth 1988) of a morphologically heterogeneous sample of crossbills from the southern Appalachians corroborated implications by previous authors that distinctive forms may use the same areas for breeding. Birds in the sample fell into two groups based on vocal differences, and the two groups were entirely separable based on multivariate analysis of morphological size characters. In univariate comparisons of bill and body size, the two forms showed moderate overlap but highly significant mean differences. The type specimens of *minor* and *pusilla* were compared to Appalachian crossbills of known vocalizations, and it was found that *pusilla* matched the larger form ("Type 2"). The type of *minor* was smaller and outside the range of variation of the total sample.

Evidence for sympatry of crossbill forms. One of the first ornithologists to suggest co-occurrence of different kinds of North American crossbills was Breninger (1894) who, working in El Paso County, Colorado, wrote that he "made an exhaustive study of the two forms of Crossbills inhabiting the vicinity" (p. 99). Monson and Phillips (1981) used one of Breninger's specimens (a nesting female) as the type specimen for their form *vividior,* described as much smaller than their larger "Class III" form, which is generally the commonest size class in most of Colorado (see Bailey et al. 1953). Griscom (1937) detailed more evidence that distinct subspecies of crossbills breed in the same regions of North America. Jollie (1953) noted that three sizes of crossbills flocked separately and nested in the vicinity of Moscow, Idaho. In Montana in the summer of 1954, Kemper (1959) found locally breeding crossbills which he interpreted as belonging to two subspecies ("*sitkensis*" and "*bendirei*"). Howell (1972) found that Central American red crossbills fell into two groups based on morphology, but the two groups were collected at different sites. Peterson (1985) also reported two or more "subspecies" breeding at the same time in New York, and Dickerman (1986b:133) wrote that "at least two and probably three subspecies have bred in the state." In summary, the evidence for overlapping breeding ranges is far more extensive and compelling than the evidence for rigid allopatry of different crossbill forms.

SUMMARY

An increase in the number of specimens compared has, for the last century, led to the recognition that North American crossbills are highly morphologically variable and may contain more than one form. Because all red crossbills together form a smooth morphological gradient, distinctions based solely on morphology have been arbitrary. The situation for *Loxia curvirostra* is different from other problem groups, because different morphotypes or "races" have not been assigned with certainty to particular breeding areas, largely due to the paucity of known breeding specimens. The lack of relevant specimens stems from the omission of gonadal information on specimen tags, and unlike many sedentary or migratory birds with defined breeding seasons and breeding ranges, the month of collection may be a weak or non-informative clue regarding breeding status in crossbills.

MATERIALS AND METHODS

SPECIMENS EXAMINED AND STUDY SITES

Specimens of known vocalizations. A majority (616) of the 720 individuals with both recorded vocalizations and known morphology (Fig. 2 and Appendix A) were preserved. Most specimens were prepared as study skins with accompanying partial skeletons, although 39 specimens were preserved only as study skins, 26 only as complete skeletons, and 25 as incomplete skeletons with flat skins. Partial skeletons included bones from the body core outward to the humerus and femur and were cut through the caudal vertebrae and orbits, leaving the bill on the skin. Incomplete skeletons included a complete skull and all major bones (excluding the right tarsometatarsus and phalanges, right wing bones, and a few distal caudal vertebrae which were left to support an accompanying flat skin). Birds not immediately preserved were either banded and released or kept in captivity for further behavioral observation.

The breeding status of specimens was assessed using behavior (nesting, pairing, singing, association with juveniles), gonadal condition, and incubation patches of females as criteria. "Breeding" females were considered those with active incubation patches, enlarged oviducts, and/or the largest ovarian follicle at least 2 mm in diameter or follicles recently ovulated; "breeding" males were those with enlarged testes (largest testis > 25 mm^3).

Museum specimens. An analysis of museum specimens (n = 2,979; Fig. 3 and Appendix B) was performed to determine how well the sample of birds with known vocalizations represented the range of morphological variation in North American red crossbills. Generally all available specimens were measured for each collection, although many important collections were not included. A subset of 520 specimens were either juveniles and/or Eurasian in origin and not used in the analysis. It is assumed here that the museum sample provides a reasonable estimate of the range of morphological variation in North American *L. curvirostra*.

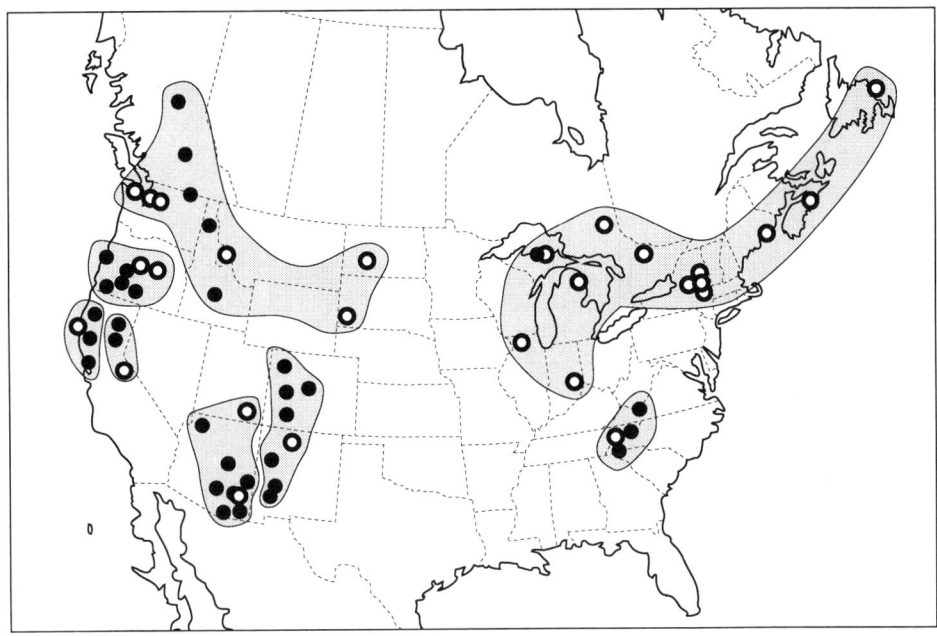

Figure 2. Recording regions and localities. Black dots indicate sites where birds were recorded and collected. Open dots show localities where crossbills were recorded by other observers. Shaded areas indicate arbitrarily defined regions defined in the text.

Age and sex determination. Museum specimens were divided into sex classes using the plumage characteristics given by Phillips (1977). Birds that are bright yellow, orange, or red are almost always males, whereas light yellow to greenish birds are almost always females. Most specimens collected for this study were also dissected for visualization of the gonads.

Young crossbills leave the nest in a brown and white streaked body plumage which is lost within the first year. For aging by plumage characters, birds were divided into ten classes based on counts of the numbers of streaked and solid contour feathers (Table 1). Subtle and often inconsistent characters such as color of flight feather edges, feather wear, or presence of buffy wingbars may distinguish some first-year from older crossbills (Phillips 1977). However, because of the unreliability of these characters, and the general observation that small songbirds reach full size within a few months of hatching (P. R. Grant 1986, T. B. Smith 1990b), different year classes of adults were lumped.

For aging by skull pneumatization, crania were examined (after they were cleaned) for the relative degree of single- and double-layering. The percentage of the skull pneumatized (double-layered) was estimated by first drawing a diagram of the skull and outlining the shapes of unpneumatized areas. The drawings were then scanned and measured using

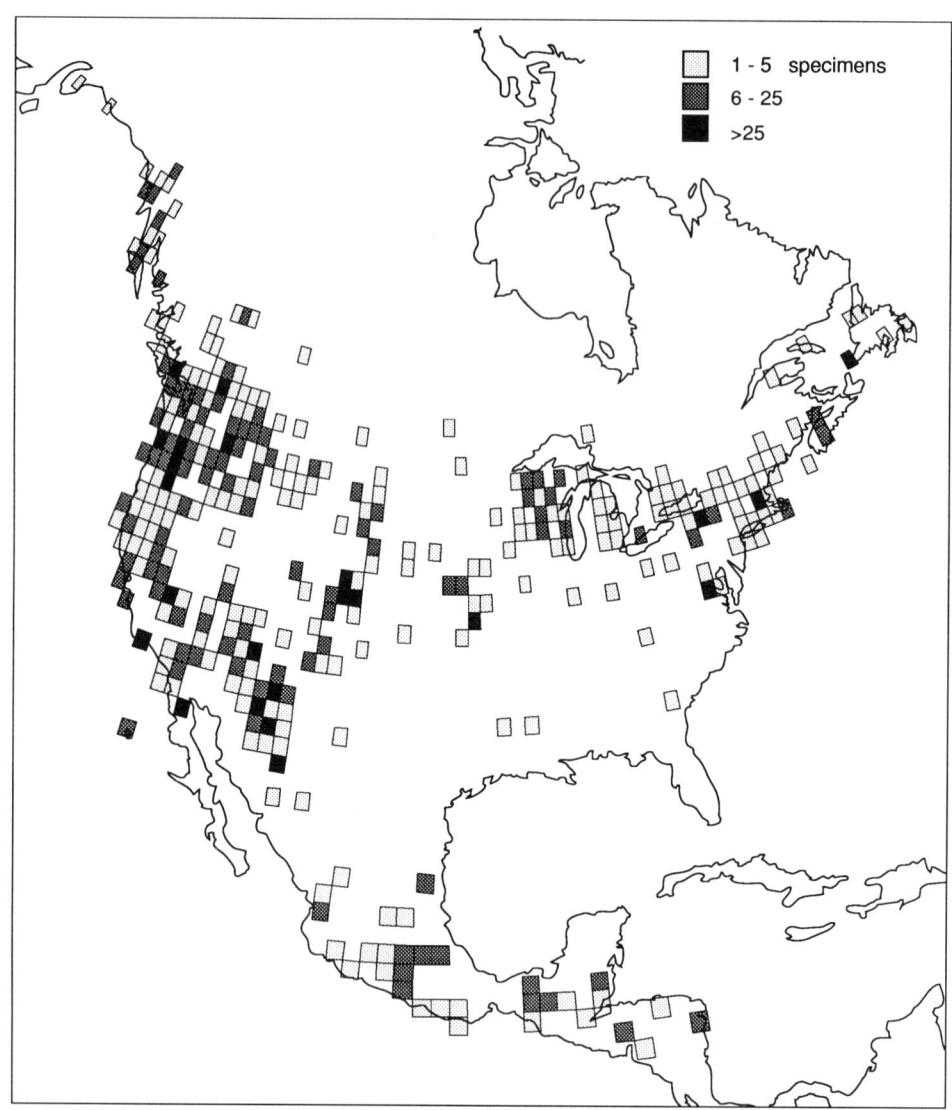

Figure 3. Distribution of museum specimens used in this study.

Table 1. Plumage Scoring Scale for Age Estimation of Juveniles and Color Description of Adult Males[a]

Score	Number of feathers
0	None
1	1–3 individual feathers
2	4–10 individual feathers
3	At least 11 individual feathers, but no more than 20% of total
4	From 21–50% of total
5	From 51–79% of total
6	At least 80% of total, but with more than 10 feathers of another category
7	4–10 feathers of another category
8	1–3 feathers of another category
9	All

a. Scores are used either for juveniles (either sex) or for adult males. For aging of juveniles by plumage, the score represents the number of non-juvenal (non-streaked) green, yellow, orange, or red body feathers. For color description of adult males, scores represent the number of body feathers in each of three color categories: red, orange, or yellow.

MorphoSys (Meacham and Duncan 1990). The single individual with the least pneumatized (yet solid enough to remain intact throughout the preparation process) skull was designated as 0% pneumatized; other individuals were scaled in a relative fashion, up to 100% pneumatized if no single-layered areas remained.

VOCALIZATIONS

Field recordings. Recordings of crossbills were accumulated in field work from 1983 to 1989. Over 100 hours of tape recordings containing either vocalizations or dialog describing crossbill behavior were made. In the early fieldwork (1983 to 1985), a reel-to-reel Uher 4000 series recorder and one of several cardioid microphones mounted in a parabola were used. Later recordings (1985 to 1989) were made with a Sony 5029A cassette recorder with two different microphone systems: (1) a Sennheiser omnidirectional microphone mounted in a Sony PBR-330 parabola, or (2) a Sennheiser shotgun microphone with a miniature in-line amplifier. There is no evidence that the different recording equipment biased the variation in vocalizations described below.

Both wild and captive crossbills were recorded to document the call repertoire. As many as 23 captive crossbills were concurrently available for study between March 1983 and 1990. Captive crossbills were vocal and facilitated field observations by attracting wild birds. The sequence of vocal development in nestlings and post-fledging young was also documented. One entire nesting sequence of a wild pair, from pre-nesting courtship to the fledging and feeding of flying young, was studied from 20 January to 21 June 1984 on Price Mountain, Montgomery County, Virginia. One captive female initiated 12 nests between 1985 and 1989, most of which fledged at least one young.

The most prominent vocalization in the repertoire of most cardueline finches is the *flight call* (Mundinger 1970, Marler and Mundinger 1975); and in red crossbills, flight calls show both individual and population distinctiveness (Nethersole-Thompson 1975; Groth 1988). To obtain calls of known individuals, birds were either recorded and then shot, or birds were first captured in mist nets and recorded when they vocalized from wire cages. Extensive recordings of individuals both before and after capture showed that captivity does not affect individually specific features of flight call structure. At least two flight calls were recorded from each individual later used in the morphological analysis. Uncounted thousands of flight calls were also recorded from other wild crossbills not captured or collected. Crossbills of known flight calls from the southern Appalachians, described in an earlier study (Groth 1988), were included. Audiospectrograms of 7,519 flight calls were prepared from the recordings of known birds, or an average of about 10 calls per individual. Thousands of other calls from these individuals were recorded but not audiospectrographed.

Other than the flight call, two other calls are frequently given by red crossbills and have shown population distinctiveness. *Alarm calls* and *excitement calls* (Groth 1988) were obtained from 95 and 31 individuals, respectively, with recorded flight calls. Many other alarm and excitement calls were obtained in the field from other (uncollected) crossbills recorded giving flight calls.

Recordings from other observers. All recordings provide tests of the hypothesis that there exist a limited number of discrete vocal groups of red crossbills. Additional recordings of crossbill vocalizations were obtained from observers in Washington, Oregon, California, Utah, Arizona, Colorado, New Mexico, Montana, South Dakota, Illinois, Michigan, Indiana, Ontario, New York, Maine, Nova Scotia, Newfoundland, North Carolina, and Tennessee (Appendix C). These recordings greatly expanded the geographic coverage for this study. Selected fragments from this set of recordings were audiospectrographed and compared to the calls of birds of known morphology. Comprehensive information on the recording equipment used by the other observers is not available.

Audiospectrography and analysis of vocalizations. Audiospectrograms were prepared on Kay Elemetrics 6061A or 7029A Sona-Graph machines. Most audiospectrograms were run at the 160–16,000 kHz frequency range using the "wide-band" setting, and all audiospectrograms illustrated in this work were graphed using these settings. Alternative displays were produced with runs at the 80–8,000 kHz frequency range or with the "narrow-band" setting. The classification phase of this project consisted of assigning individuals to groups on the basis of similarity in vocalizations. For this process, the single clearest audiospectrogram of a flight call was used to characterize each individual. Birds were divided into groups based on visual inspection of similarity in flight call structures. A second phase of classification included the inspection of variation in alarm and excitement calls from birds of known flight calls. The latitude in flight call variation within each hypothetical group was determined in part by the correspondence among these three major calls. The extent to which distinctive variants in these other calls were predicted by flight call structure was examined.

MORPHOLOGY

Field measurements. Nine morphological characters were taken immediately following capture or collection of specimens in the field. Body mass was measured with a double-beam or digital balance to the nearest 0.1 gm. Linear measurements were taken with dial or digital calipers to the nearest 0.1 mm as follows: wing length—chord of the right wing supported about 5 mm from the body; tarsus length—distance from the posterior surface of the bent tibiotarsal-tarsometatarsal joint to the distal edge of the lowest undivided scute on the right leg; upper mandible length—distance from the anterior edge of the nostril to the tip; lower mandible length—distance from the juncture of the lateral rami of the lower mandible to the tip; bill depth—taken with the bill fully compressed as a line perpendicular to the nostrils; upper mandible depth—taken with the bill open as a line perpendicular to the nostrils; upper mandible width—the distance between tomia at a plane intersecting the nostrils; lower mandible width—distance between outer edges of the rami of the lower mandible where they enter the skin.

Study skin measurements. Nine measurements were taken from study skins, five of them in a way identical to those taken in the field: upper mandible length, lower mandible length, bill depth, upper mandible width, and lower mandible width. The sixth bill character, diagonal upper mandible depth, was taken as the distance between the upper edge of the culmen above the center of the nostril to the nearest point on the edge of the tomium on the left side of the upper mandible. Wing length was measured with the wing flat against the body. Tail length was measured from the point of insertion of the central rectrices to the tip of the longest rectrix. Toe length was measured on the third digit of either the left or right foot, from the proximal edge of the scute covering the joint between the basal and medial phalanges to the point of insertion of the claw.

Populations and sibling species of birds often differ in wing shape, and relative feather lengths may be diagnostic (e.g., N. K. Johnson 1963). On the specimens collected for this study, the lengths of the outer seven primary feathers of the right wing were taken with calipers to the nearest 0.1 mm with the wing flat against the body. Shapes of wings were compared by plotting wing profile diagrams.

In addition to size-related characters, the relative curvature and shape of the bill in crossbills might predict specific or subspecific limits (e.g., Griscom 1937, Knox 1976). The culmen of crossbills appears to curve in a nearly circular fashion, such that indices of curvature based on the fraction of a circle traversed by the bill seemed appropriate. The *curvature angle* developed for this study (Fig. 4) is proportional to the ratio of the distance from the midpoint of the chord between points at the base and the tip of the bill and the height of the corresponding arc (edge of the culmen). Measurements were taken by placing study skins (specimens with recorded vocalizations only) under a video camera and

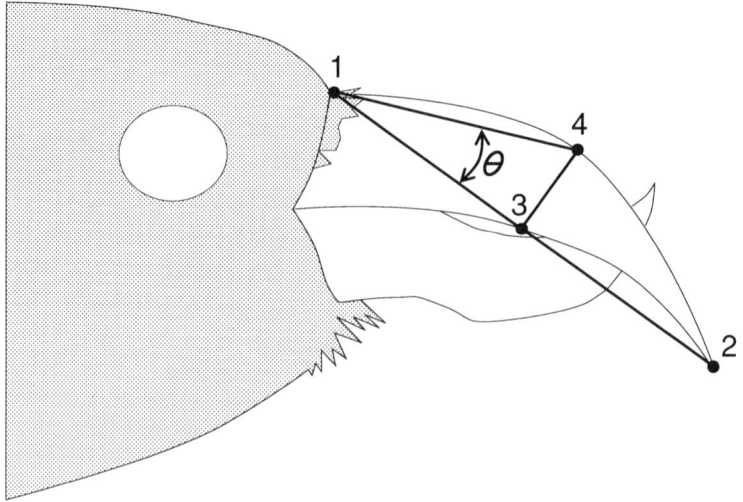

Figure 4. Method of bill curvature measurement. Points 1 and 2 were taken at the base and tip of the culmen. Point 3 is the midpoint between points 1 and 2, and point 4 is the intersection by the culmen of a line perpendicular to line 1–2 at point 3. The curvature angle θ was taken as shown.

digitizing the profile of the head and bill on a computer screen using MorphoSys (Meacham and Duncan 1990).

Skeletal measurements. Twenty-two linear measurements were taken on skeletons using dial or digital calipers. Most measurements were taken to the nearest 0.1 mm, with the exceptions of coracoid, humerus, femur, and tibiotarsus widths, which were taken to the nearest 0.01 mm. Measurements generally matched those for other songbirds illustrated by Robins and Schnell (1971), with these differences: skull width was taken as the maximum width of the skull between the outer edges of the zygomatic processes (using terminology of Richards and Bock 1973); skull depth was taken with one jaw of the caliper flat against the base of the cranium to the most distant point on top of the skull; synsacrum width was taken as the distance between the outer edges of the pectineal processes; anterior synsacrum length was taken as the distance between the anterior edge of the ilioischiatic fenestra to the most anterior projection on the synsacrum; finally, coracoid, humerus, femur, and tibiotarsus widths were taken at their minima by successively rotating the bones and repeating the measurement several times at different positions.

Differences between right and left legs within individuals, or "handedness," a phenomenon described by Knox (1983) as associated with bill-crossing direction, was considered a potential source of intrapopulation variability. Knox (1983) studied only tarsus length on study skins and not measurements of leg bones. To evaluate the impact of handedness on the variance in leg measurements, 18 measurements (9 characters per leg) were taken on 38 skeletons of *Loxia* with known bill-crossing direction in the Museum of Vertebrate Zoology. The characters were the lengths, minimum widths, and distal end widths of the femurs, tibiotarsi, and tarsometatarsi (see above; also Robins and Schnell 1971). Measurements were taken with digital calipers to the nearest 0.01 mm. The absolute differences between right and left sides were calculated, and the significance of the association between crossing direction and bone size was tested using two-tailed paired t-tests. The tests were performed using SAS PROC MEANS (SAS Institute 1985:799) comparing the mean differences between the two legs to zero.

Comparisons between museum specimens and new specimens collected in this study. The range of morphological variation in the museum specimens was summarized using principal components analysis (PCA). Raw data were input into SAS (1985) PROC PRINCOMP and the principal components extracted from the correlation matrix among 8 of the 9 characters (bill depth was excluded because of variation in degree of separation of the mandibles due to preparation technique). Individuals of known vocalizations were given scores along the principal components derived from the museum specimens using SAS (1985) PROC SCORE by transformation using the eigenvectors for each character. Scores for individuals were examined on scatter diagrams of the first three principal components.

Morphological analysis of crossbills of known vocalizations. Specimens with recorded vocalizations were separated into groups on the basis of call note similarity, not geographic origin. Birds within these hypothetical groupings were divided further by age and sex, because juveniles and females tend to average smaller than adult males in most kinds of songbirds. Although crossbills in streaked plumage may nest and function as full adults in wild populations, "adults" were here conservatively defined for morphological purposes as birds that had lost all streaked juvenal body feathers (age class 9, Table 1). For these samples, the mean, standard error of the mean (SE), range of variation, and coefficient of variation (CV) were computed for each character. Levels of the significance of mean differences between samples were estimated using two-tailed t- or F-tests. Sexual dimorphism in univariate characters was calculated using ratios of male means over female means and calculating the percentage differences. The potential contribution of age to variation in size was examined by comparing measurements of juveniles at various stages of development to within-group means for adults.

For multivariate analysis of the skeleton, missing values on slightly damaged specimens were replaced with predicted values based on linear regressions with the most highly correlated characters in the pooled sample of all adults (separated by sex). This procedure has the advantage over insertion of group means for missing values in that it will not exaggerate differences between groups; however, slight decreases in pooled-sample variance may result. Replacement of missing values was assumed to be of minimal impact because less than one percent of all data was replaced. If a specimen showed more than three missing values it was not used beyond the calculation of univariate statistics. Sclerotic ring width was excluded from multivariate analysis due to an excessive number of missing values.

Principal components analysis (PCA) was chosen as a method for exploratory ordination. Morphological characters were divided into two major complexes for PCAs: (1) bill size characters, consisting of the 6 bill measurements taken in the field, and (2) body size characters, consisting of 21 skeletal measurements (sclerotic ring width was excluded, due to missing values). For these analyses, it was not assumed that the sample consisted of more than a single group. One problem with this method is that loadings on principal components may be confounded by within-group trends if the sample contains more than one group (James and McCulloch 1990), whereas PCAs on population means give appropriate data reduction. The PCAs in this study were not influenced by *a priori* information about group structure, so that mapping of hypothetical groups on PCA plots might reflect a less biased separation among individuals.

Canonical discriminant analysis (PROC CANDISC; SAS 1985) was chosen as an *a posteriori* method to find optimal linear combinations of variables separating groups and to measure relative morphological separation among groups. This method first transforms variables so that the pooled within-group covariance matrix is an identity matrix, and then it performs a principal components analysis on the means to maximize the variance among groups. The resulting principal components are termed *canonical axes,* with eigenvalues summarizing the ratio of between-group to within-group variation on each axis. One analysis was performed to measure the overall morphological separabilty of groups using 7 study skin measurements. These analyses produced coefficients that allow estimation of group membership for unclassified specimens, such as museum study skins. A larger analysis using a combined data set with 32 morphological characters (including all those listed in Appendix D, and excluding diagonal upper mandible depth, sclerotic ring width, and bill curvature angle) was used to calculate Mahalanobis' distances (D^2; Sneath and Sokal 1973) between group centroids, which were used as generalized estimates of relative morphological distance among groups.

To examine geographic differences in morphology within vocally defined groups, a single group with a broad geographic occurrence and large sample size was divided into geographic subsamples. PCAs using the correlation matrices of raw measurements for the two sexes were used to create multivariate morphospaces in which to visualize geographic trends. Because many of the specimens in this group lacked skeletal data, only the 9 field measurements were used.

PLUMAGE COLOR DESCRIPTION

Standard reflectance spectrophotometry may not be appropriate in color analysis of this group for two reasons: first, pigmentation generally occurs only at the tips of the body feathers, and in most skins the grayish bases of the feathers would interfere with colorimetric readings; second, few specimens of males appear solid in color. In order to describe total body coloration, a new color–coding system was designed for crossbills that used a visual count of the number of yellow, orange, and red body feathers. *Yellow* feathers were classified as those, using the terminology in Ridgway (1912), ranging from aniline yellow and yellow ochre to any shade with a more greenish hue. *Orange* feathers ranged from cadmium yellow through xanthine orange and bittersweet orange. *Red* feathers ranged from peach red through nopal red, Brazil red, scarlet, and flame scarlet. Counts were made of the number of feathers of each of the three major colors, and specimens were given three-digit codes for yellow, orange, and red, respectively (Table 1). For example, a specimen with a score of 207 had from 3 to 10 individual yellow feathers (score of 2 for yellow), no orange feathers (score of 0 for orange), and the remainder red (score of 7 for red). A fully red bird of any shade would be 009. This system allocates specimens to one of 88 potential character states. Groups of crossbills were compared with respect to the distribution of specimens among these states.

ALLOZYME ELECTROPHORESIS

Tissue preparation. Within two hours of collection, the entire liver and heart, most of the kidneys, and a piece of pectoral muscle were packed in a cryogenic tube and frozen in liquid nitrogen and later transferred to an ultracold freezer at -70° C. For each individual, small pieces (about 50 mg) of the four tissue types were mixed together and ground with a razor blade or rotary tissue grinder in an equal volume of deionized water at 0–10° C. The slurry was centrifuged at 10,000×g for 30 min. at 3° C and the aqueous tissue extract was retained and stored at -70°.

Gel electrophoresis. Gel slabs were made of 12.5% starch with one of four different buffer solutions (recipes given in Selander et al. 1971) and assayed (following Harris and Hopkinson 1976) for 30 protein systems corresponding to 43 presumptive genetic loci (abbreviations follow Richardson et al. 1986, except that the peptidases are abbreviated according to their substrates) as follows: LA [leucyl-alanine], LGG [leucyl-glycyl-glycine], and PAP [phenylalanyl-proline]): (1) LiOH buffer for LA, LGG, PAP, ACP, EST, ALB, HB, and GDA; (2) Poulik buffer for LDH, GPI, and CK; (3) TC8 buffer for ICD, PGM, α-GPD, NP, SDH, ME, GLUD, GOT, MPI, EAP, GPT, MDH, ALD, G-6-PDH, ADH, and ADA; and (4) TM buffer for ACON, 6-PGD, and SOD. Richardson et al. (1986) contains Enzyme Commission numbers for these proteins. Six loci (ACON-2, ADH, ADA, PGM-2, G-6-PDH, and ME) were resolved inconsistently and not used in the analysis. General protein bands representing possible genetic loci ("ALB-2" and "ALB-3") were scored but not used in the final analysis, because allelic homology is questionable with non-specific assays. Bands at LA-2 were difficult to resolve because of their similar mobilities to bands representing the most common alleles at LA-1, so long runs at high temperatures were performed to selectively denature LA-1 (the dimer) without adversely affecting LA-2 (the monomer).

Electrophoresis was repeated if samples showed uncommon heterozygotes or ambiguous patterns on the first run. Alleles at each locus were labeled according to measurements of their mobilities relative to the most common allele, which was designated *100* (or *-100* in cathodally migrating loci), and individuals with low-frequency alleles were repeatedly run side by side for calibration.

Samples and analysis. Samples of 561 *L. curvirostra* and 16 *L. leucoptera* were electrophoresed. The *leucoptera* were collected in widely separated areas of North America, including Alaska (n = 3; these were the same individuals used previously by Marten and Johnson 1986), British Columbia (n = 2), Colorado (n = 4), and New York (n = 7).

All calculations with the electrophoretic data set were performed using BIOSYS-1 (Swofford and Selander 1981). Genetic structure within each of the two major samples (*curvirostra* and *leucoptera*) was examined by calculation of the fixation index or inbreeding coefficient (*F*), which provides a measure of deviation from Hardy-Weinberg predictions (Wallace 1981) treating the entire sample as one population. Significances of *F* values within loci were tested with chi-square goodness-of-fit tests by pooling genotypes into three classes at each locus: (1) homozygotes for the commonest allele, (2) heterozygotes with the commonest allele and another allele, and (3) all other genotypes. This pooling of genotypes avoids the problem of sampling error over genotypes when some are rare (Swofford and Selander 1981). The *curvirostra* were further divided into subsamples based on vocalizations (see above), which were considered "demes" in this study. Genetic structure within and among demes was analyzed using *F*-statistics (Wright 1951, 1965). The among-population component of variation was calculated in two ways: with Nei's (1977) multiallelic version of F_{ST} or "G_{ST}" using STEP FSTAT; and with Wright's (1978) F_{ST} using STEP WRIGHT78, which accounts for sampling error of alleles within populations. The latter generally provides a more conservative estimate, and these values were tested for significance using Workman and Niswander's (1970) formula for c^2 of where $c^2 = (k - 1)(2N_T F_{ST} - 1)$ in which k is the number of alleles and N_T is the number of individuals. The test has $(k - 1)(s - 1)$ degrees of freedom, where s is the number of populations.

Gene flow among samples was estimated according to Slatkin's (1981, 1985a) approach using private alleles. This method is based on the relationship between the mean frequency of alleles found in only one deme (unique alleles) and the quantity Nm, the average number of migrants between demes per generation, where N is the number of individuals per deme and m is the migration rate between demes. The formula used was: $\ln(p) \approx -0.505\ln(Nm) - 2.440$, where p is the average frequency of alleles found in only one sample. Corrections were made by dividing the average sample size by 25 and dividing estimates of Nm by the resulting value (Slatkin 1985a, Slatkin and Barton 1989). An estimation of Nm between *leucoptera* and *curvirostra* was calculated as a basis of comparison for the significance of Nm estimates among "demes" within *curvirostra*.

Genetic distances between samples were calculated following Rogers (1972) and Nei (1978). Nei's distances have been widely published for other groups and provide a means of comparison across taxa. Rogers' distances satisfy the triangle inequality (Sneath and Sokal 1973, Wright 1978, Rogers 1986) and were therefore chosen for construction of phenograms portraying patterns of genetic distance. Plots were calculated using the distance Wagner approach (Farris 1972, 1981; Swofford 1981) with rooting of the tree at *leucoptera*, and multidimensional scaling (MDS; Kruskal 1964, Sneath and Sokal 1973) which produces ordinations of reduced dimensionality with best fit to original distance matrices. The MDS routines were run for three dimensions, using MDSCALE of NTSYS (Rohlf et al. 1982) and programmed to begin with an initial "tree" configuration using the operation TAXON and to stop iterating at a minimum stress (Kruskal 1964) value of 0.001 or at 50 iterations, whichever came first. A minimum spanning tree was superimposed on the MDS plot to check for distortions, and the quality of this plot was assessed using the cophenetic correlation coefficient (r_{cc}; Sneath and Sokal 1973) between Rogers' (1972) distances and observed distances in the plot.

The patterns of genetic relationship among call types were compared to differences based on morphology and vocalizations. Trends of correspondence between Rogers' (1972) genetic distances and Mahalanobis' distances in morphology (see above) were contrasted visually in a scatterplot. The hypothesis that genetic and morphological distance matrices were independent (null model) was tested using the Mantel test (Mantel 1967, Schnell et al. 1985) which, if falsified, means that the matrices share a common structure. This test results in a *t*-statistic to be compared against the standard normal distribution (infinite degrees of freedom). The genetic distances were computed using specimens of all ages and both sexes, whereas the morphological distances were from samples of adult males only.

COMPARISON OF MUSEUM SPECIMENS OF CROSSBILLS TO SPECIMENS OF KNOWN VOCALIZATIONS

Correlation coefficients among morphological characters of museum specimens were high (Table 2), and all loaded highly and positively on the first principal components (Table 3), showing that much of the observed variation in both sexes can be interpreted as "size." PC1 vectors accounted for approximately 80% of the sum of the standardized variances in 8 characters for both sexes. Toe length had the weakest correlations with the other variables (which may indicate error due to degree of bending of toes on study skins) and was the highest loading variable on PC2. PC3 axes in analyses of both sexes had wing and tail lengths loading opposite in sign to the 5 bill characters, allowing the interpretation that PC3 axes were indices of relative bill size versus "body size" as measured by wing and tail lengths. For this reason PC3 was chosen over PC2 in two-dimensional plots so that variation in toe length would not bias interpretation. Although PC3 vectors for the two sexes were similar, they were inverse in sign; in the PCA for males, individuals with relatively large bills and short wings and tails obtained more negative PC3 scores, and in females such individuals were more highly positive.

Distributions for both sexes contained greater numbers of points at PC1 values between 0 and +2 (Fig. 5), indicating a greater representation of medium-sized to large individuals. The clusters, especially the one for males, appear to be comprised of four interconnected smaller clusters separated along PC1. The most diffuse regions of the clusters were at the far positive ends of the PC1 axes in morphospace containing the largest specimens. The occurrence of four modes along PC1 is consistent with previous arguments (Phillips 1977, Monson and Phillips 1981) for a four-group division of the complex based on overall size.

Samples of males and females of known vocalizations (Fig. 5) nearly spanned the range of PC1 values of museum specimens, indicating that natural size variation in the complex was fairly well represented. Some "shape" variation among museum specimens, as measured by scores on PC3 (highly negative in males, positive in females), was not evident among the samples of birds of known vocalizations. Further investigation showed that large-billed birds with short wings and tails, confined to Central America, were

Table 2. Correlation Coefficients[a] Among Characters in the Sample of Museum Study Skins of North American Adult Red Crossbills

	Characters							
	1	2	3	4	5	6	7	8
1. Toe length	—	0.59	0.67	0.56	0.59	0.66	0.63	0.68
2. Tail length	0.56	—	0.87	0.70	0.69	0.72	0.69	0.74
3. Wing length	0.67	0.88	—	0.77	0.76	0.81	0.77	0.82
4. Upper mandible length	0.59	0.76	0.81	—	0.90	0.78	0.76	0.81
5. Lower mandible length	0.61	0.75	0.81	0.91	—	0.76	0.73	0.79
6. Upper mandible depth	0.67	0.75	0.84	0.82	0.81	—	0.86	0.89
7. Upper mandible width	0.66	0.73	0.82	0.80	0.80	0.88	—	0.88
8. Lower mandible width	0.68	0.76	0.84	0.83	0.83	0.89	0.89	—

a. Values for females (n = 834) above the diagonal; males (n = 1,513) below. All values are significant at $P < 0.001$.

Table 3. Correlations (Factor Loadings) Between Study Skin Measurements and the First 3 Principal Components for Samples of Museum Specimens of North American Red Crossbills

Variable	Males			Females		
	PC1	PC2	PC3	PC1	PC2	PC3
Toe length	0.751	0.634	0.082	0.754	0.570	0.173
Tail length	0.864	-0.197	0.430	0.851	0.065	-0.485
Wing length	0.930	-0.062	0.250	0.918	0.070	-0.282
Upper mandible length	0.910	-0.193	-0.164	0.892	-0.325	0.078
Lower mandible length	0.912	-0.159	-0.170	0.882	-0.290	0.073
Upper mandible depth	0.933	0.018	-0.114	0.918	0.006	0.131
Upper mandible width	0.922	0.034	-0.153	0.899	-0.003	0.192
Lower mandible width	0.940	0.027	-0.122	0.934	-0.006	0.122
Total variance[a]	80.5%	6.4%	4.5%	78.0%	6.5%	5.3%

a. Percentages of total standardized variance explained by each axis.

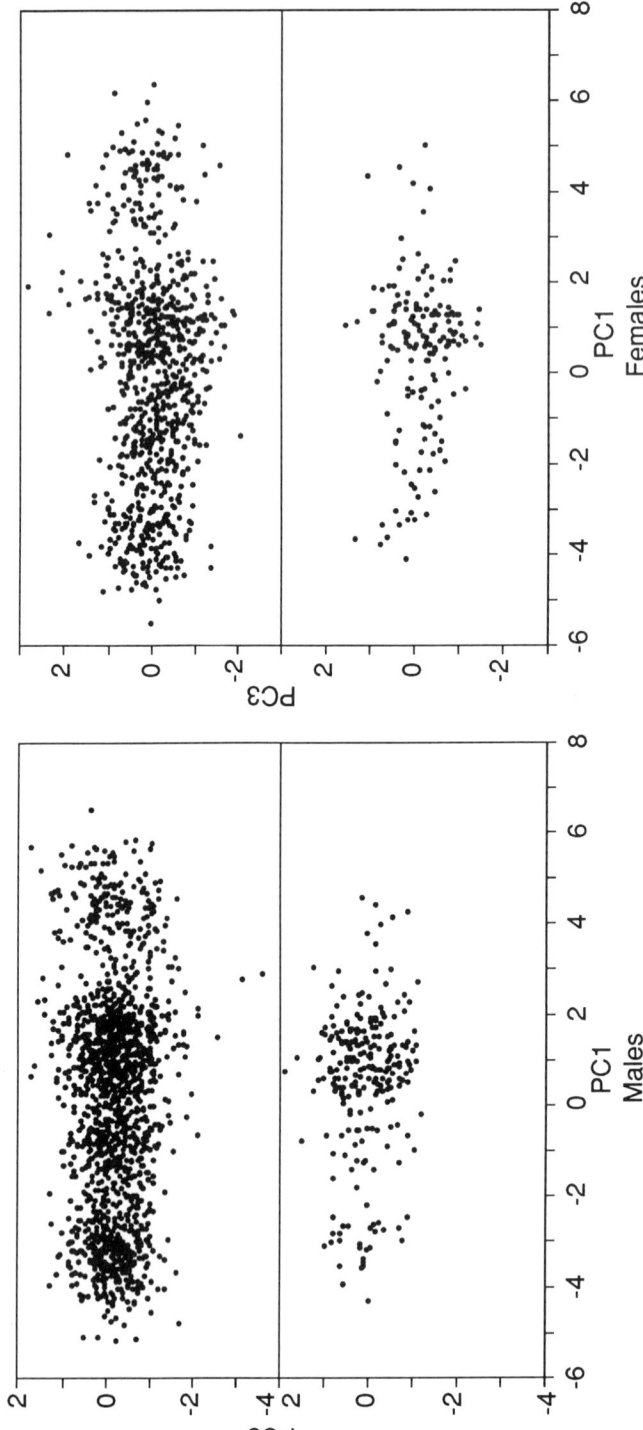

Figure 5. Principal components plots based on 8 characters of museum study skins of North American *L. curvirostra*. Upper plots show the distributions for the specimens from which the principal components were derived; lower plots show distributions in the same morphospaces for the study skins of crossbills of known vocalizations.

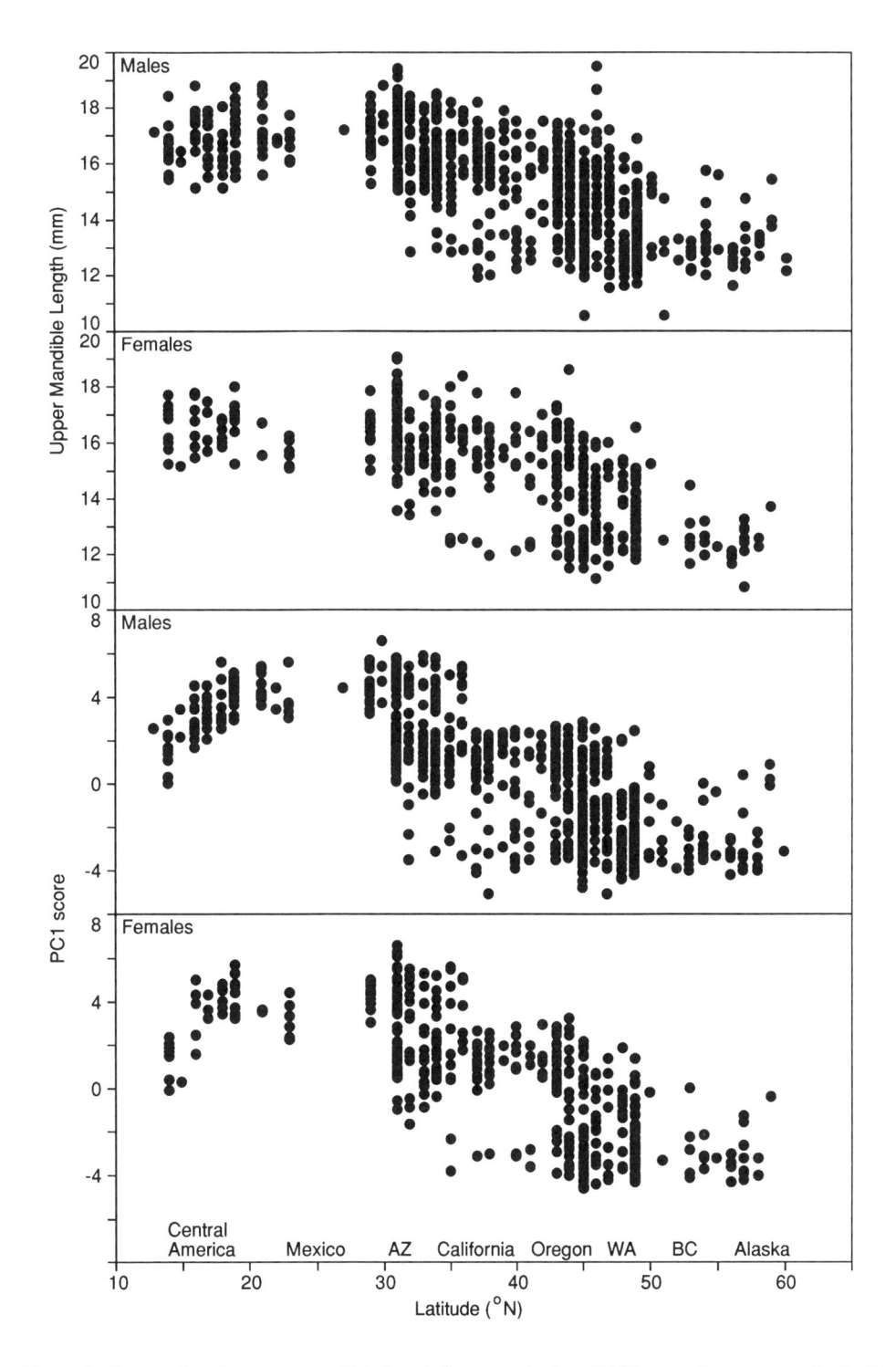

Figure 6. Scatter plots for upper mandible length (two top plots) and PC1 score (bottom two plots) with latitude in *L. curvirostra* collected along a western transect. Note the trend for decreasing size with increasing latitude.

Table 4. Clinal Variation in Museum Specimens of North American Red Crossbills Collected Along a Western Transect

Variable	Males			Females		
	N	r[a]	slope (SE)[b]	N	r[a]	slope (SE)[b]
Toe length	927	-0.592	-1.538 (0.069)	482	-0.549	-1.498 (0.104)
Tail length	920	-0.590	-1.783 (0.081)	478	-0.562	-1.812 (0.122)
Wing length	925	-0.670	-1.814 (0.066)	481	-0.631	-1.717 (0.097)
Upper mandible length	915	-0.680	-3.649 (0.130)	480	-0.662	-3.660 (0.189)
Lower mandible length	915	-0.679	-3.656 (0.131)	480	-0.653	-3.594 (0.191)
Bill depth	710	-0.717	-3.290 (0.120)	370	-0.685	-3.178 (0.176)
Upper mandible depth	920	-0.739	-3.480 (0.105)	479	-0.711	-3.227 (0.146)
Upper mandible width	921	-0.722	-3.476 (0.110)	481	-0.698	-3.267 (0.153)
Lower mandible width	908	-0.736	-3.621 (0.111)	477	-0.699	-3.375 (0.159)
PC1	884	-0.743	—	465	-0.718	—
PC2	884	-0.093	—	465	0.021	—
PC3	884	0.139	—	465	-0.115	—

a. Product-moment correlation coefficients between latitude and raw measurement.
b. Slope of $\log_{10}$-transformed values regressed on latitude; slopes and SEs multiplied by 1000.

responsible for the discrepancy. Therefore, the evolutionary patterns described in this monograph are limited to northern North America. Other than this difference, clouds of points for individuals of known vocalizations were like "skeletons" of the clouds for museum specimens, and like the museum samples had greatest scatterplot densities in the medium to large size range.

Variation in North American *L. curvirostra* shows a general pattern opposite the predictions of Bergmann's rule. Along a western transect (all specimens from Alaska, British Columbia, Washington, Oregon, California, Arizona, Mexico, Guatemala, Belize, Honduras, and Nicaragua) there exists a cline in size with smaller birds in the north (Fig. 6). The smallest crossbills, with PC1 scores around -4, were collected from Alaska south to Arizona, including some of the same latitudes occupied by the largest birds. The plots of PC1 scores for both males and females show that the largest crossbills occur in Mexico, but overall size decreases somewhat in Central America. In contrast, upper mandible length does not decrease in Central America, indicating that crossbills at the southern end of the range have longer bills relative to overall size. Much of the morphological variation in the complex was contained within narrow spans of latitude. For example, males at the extremes of bill length (Fig. 6, top) were collected between 43° and 47° N latitude, where one male had a bill nearly 100% longer than another.

Unfortunately, it was not possible to identify local breeding populations of crossbills because gonadal information was rarely included on specimen tags. At least for western North America, July appears to be the peak month for crossbill breeding (Groth, unpubl. data). Separate analyses (not shown) of the same western transect, using only crossbills taken in June, July, and August, showed trends nearly identical to those in the pooled samples. This suggests that using specimens from all months did not obscure the overall pattern.

All characters showed similar trends with latitude (Table 4), but bill size characters showed greater rates of change (as indicated by slope values) than toe, wing and tail. PC1 scores for both sexes were more highly correlated with latitude than univariate characters. Scores along PC3 for both sexes showed low but significant correlations with latitude; this may reflect either the presence of crossbills with large bills and short wings and tails in Central America, or allometric trends in which large-billed birds have proportionately smaller bodies.

THE VOCAL REPERTOIRE

Bird vocalizations can either form discrete categories or can be *graded* (sensu Jenni et al. 1974, E. H. Miller 1979, Maier 1982, Morton 1982). Species-specific aspects of vocalizations generally exhibit discrete differences in sequences and structures of elements (Marler 1957, Emlen 1972, Shiovitz and Lemon 1980, Becker 1982, MacNally et al. 1986); therefore, discrete vocalizations will be of most interest in systematic comparisons of crossbill populations. Furthermore, bird vocalizations are often divided into calls and songs. Calls of crossbills are defined here as any vocalizations that are not song; song is defined as notes of different types connected in a rhythmic or repetitious series. Both sexes of red crossbill were observed singing songs, but singing was more frequent and often louder in males. The complexity of song in crossbills prevented full analysis in this work. The documentation of the vocal repertoire is largely confined here to the Type 2 group of red crossbill, one of two southern Appalachian forms documented previously (Groth 1988), which was extensively recorded for this study. Several other vocalizations were less frequent than those in the basic repertoire described here (see Groth 1990 for details).

CALLS OF ADULTS

Flight call. Flight calls, the most frequent vocalization in crossbills, are given by both sexes while in flight and while perched. Coutlee (1971) used the term "contact call" for the same vocalization in three species of North American goldfinches (*Spinus=Carduelis*), and Nethersole-Thompson (1975) used *"chip"* for this call in both the Scottish (*L. scotica*) and common (*L. curvirostra*) crossbills. The familiar "kip-kip-kip" or "jip" used in popular field guides is equivalent to the flight call. Coutlee (1971) reported that the calls of goldfinches were slightly lower in pitch in perched birds (but not different in general structure), but this was not observed in crossbills.

The sequences of flight calls from two different individuals (Figs. 7A and 7B) are similar in having downward frequency modulation in a main element between 2 and 4 kHz, a feature that defines the "Type 2" grouping (Groth 1988). The calls in Fig. 7B differ from those in Fig. 7A in the structure of the main element, and they also have a non-harmonic

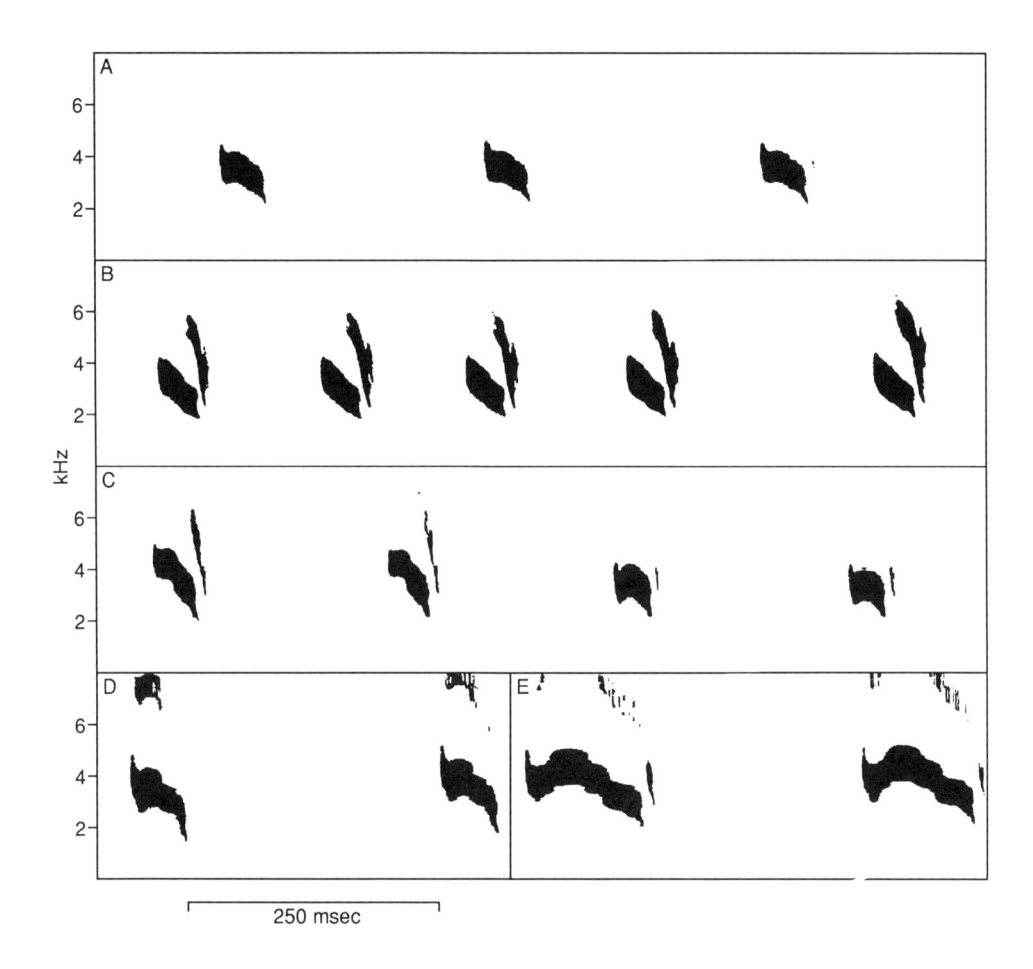

Figure 7. Flight calls of Type 2 crossbills. (A) Typical flight call series [aM413]. (B) Example of a rapid series of flight calls by aM593. (C) "Bilingualism" in aM415 [note structural differences between first two and second two calls]. (D) Typical flight calls of aM332. (E) "Alternate Type 2" calls of the same individual, aM332.

second element of higher frequency and steeper modulation beginning at the midpoint (in time) of the main lower element.

Flight calls were given at varying rates. One common cadence is shown in Fig. 7A, in which calls are given in triplicate with the notes separated by 200–250 msec and the triples more than 800 msec apart. Rate of flight call production seemed to be a general index of an individual's level of overall "excitement." A sudden new encounter with the calls of wild crossbills in the field was generally instantly followed by loud, rapid series of flight calls by captives. Calls would increase to a rate of up to 6 or 7 per second under some conditions. When in a less excited state, captives and wild birds occasionally gave single or double calls separated by one or more seconds. Arenas of singly caged crossbills became more or less vocal in synchronous responses to one another. Flight calls varied greatly in amplitude (loudness) within individuals, but neither rate of calling nor amplitude had much influence on pitch or structure evident on audiospectrograms.

Long periods of captivity had little effect on flight call structure within individuals. However, there is evidence that adults have the capacity to imitate flight call structures of other individuals under certain circumstances, as has been observed in other cardueline finches (Mundinger 1970, 1979). Mated pairs of crossbills match each other extremely precisely in flight call structure (Groth 1993), as has been observed in Cassin's finches (*Carpodacus cassinii*; Samson 1978) and various other carduelines (Mundinger 1970, 1979), suggesting that at least one member of the pair is copying the other. In rare instances, individuals were observed to give flight calls of two distinct structures on the same day and even in the same series (Fig. 7C); however, among all crossbills recorded in this study, less than ten showed this type of "bilingualism."

Some western Type 2 birds gave an unusual flight call, which was termed the "alternate Type 2," in addition to a typical and more frequently used flight call (Figs. 7D and 7E). These calls were similar to typical Type 2 flight calls in some aspects of shape and pitch, but were much longer in duration and sounded somewhat like the whistled calls of the evening grosbeak (*Coccothraustes vespertinus*). The alternate Type 2 calls were given in series of one to three and were always intermixed with "typical" calls by the same bird, but the two calls never graded together. This call was heard in only five individuals: four near Chiloquin, Oregon, in June 1985 (Appendix A) of which two (aM328 and aM332; see Fig. 15) were collected; the fifth bird heard giving this call was aM462 from northeast Washington in July 1986 (Appendix A and Fig. 14). No long-duration variant flight calls were observed in other groups (call types) of crossbills.

Distress call. Crossbills generally did not vocalize when hanging in mist nets. While handled (which may simulate capture by a predator), many individuals gave harsh, screeching cries as they struggled. These *distress calls* covered a broad frequency spectrum (Fig. 8A), with the dominant frequency generally at 6–8 kHz in Type 2 birds. Durations of these calls and the pauses between them were highly variable within individuals, and few consistent differences among individuals were apparent. Many other kinds of birds give this kind of squealing call when handled. Coutlee (1971) described distress cries of similar structure in goldfinches.

Alarm call. Nethersole-Thompson (1975) did not describe an alarm call for the Scottish crossbill. However, he described an "almost inaudible *ooks-ooks*" (p. 124) from a female he had lifted off a nest of eggs. Crossbills held in the hand not only gave distress calls (described above), but many gave softer, less harsh, shorter-duration calls (alarm calls), which were often given in graded series with distress calls (Fig. 8B). Most often, alarm calls were given singly or in series with no grading into other calls (Fig. 8C). On many occasions when hawks or corvids appeared near the field study sites, the caged decoys or other wild crossbills became silent or gave alarm calls. Alarm calls of captives or

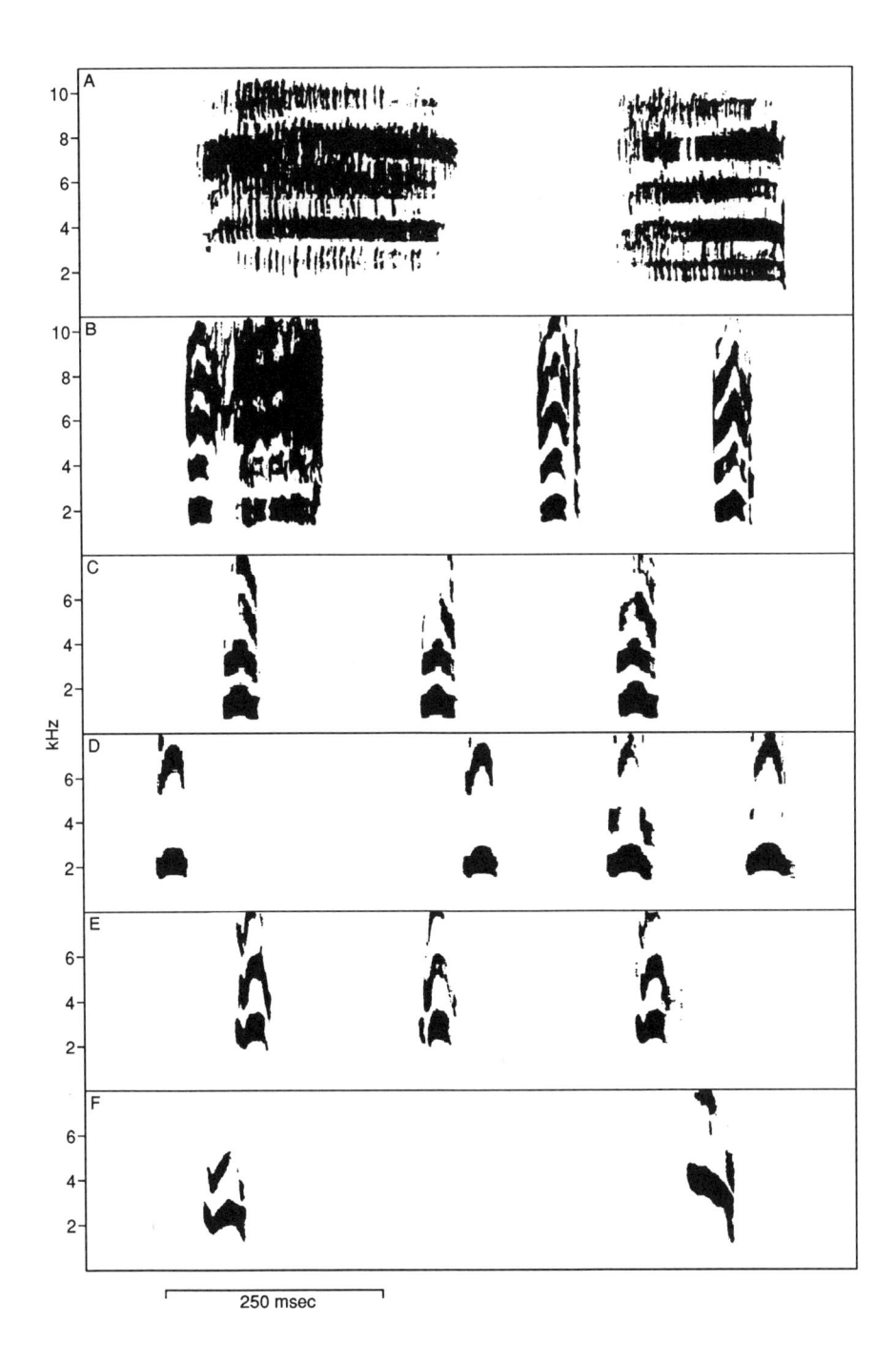

Figure 8. Distress, alarm, and excitement calls of Type 2 crossbills. (A) Distress calls of aF562. (B) Graded series of distress to alarm calls by aM680. (C) Typical series of alarm calls of aM335. (D) Grading between alarm and excitement calls (note addition of bands at 3–4 kHz in the third note). (E) Typical series of excitement calls by aF381. (F) Lack of grading between excitement calls (note 1) and flight calls (note 2) in aM20.

wild birds showed no structural differences from those given by birds in the hand, except that grading into distress calls was observed only in handled birds. Alarm calling by wild crossbills or captives was often accompanied by a stereotyped posture including elevation of the body from the perch and extension of the head and neck with the bill at a slightly upright tilt. No alarm calls were ever heard from flying crossbills. When alarm calls were heard from caged crossbills, search in the sky often revealed a large bird.

Excitement call. Nethersole-Thompson (1975) described a "deep *toop*" of the Scottish crossbill that he regarded as the most important of its calls. He wrote that birds "*toop* in alarm or anger, or excitement before flying to the nest to feed brooding hens; sometimes in flight, and in a wide spectrum of emotional situations in flock or group" (p. 121). Crossbills from North America gave distinctive calls under the same behavioral circumstances, here termed excitement calls. Nethersole-Thompson (1975:121) found that "the hen's rendering of this call is a more metallic but explosive *zoop* which is separable from that of the cock"; however, his audiospectrogram showing both sexes (p. 130) indicates no difference between the sexes. No sexual differences in excitement calls were evident in North American crossbills. Excitement calls, structurally similar to alarm calls, were louder and slightly higher in pitch. Alarm calls of Type 2 birds generally lacked the initial frequency drop contained in excitement calls. In this group, excitement calls emphasized an upper harmonic element only during the initial frequency drop of the notes; in contrast, alarm calls always had most of the sound energy in the lower fundamental frequency. Occasionally alarm and excitement calls graded together within a calling bout (Fig. 8D). Excitement calls were typically given with a cadence similar to that for flight calls (Fig. 8E). No grading between flight calls and excitement calls was noted. In all instances in which a crossbill gave the two calls within the span of a second or two, there was an abrupt transition between the two categories of calls (Fig. 8F).

Nethersole-Thompson (1975:131) also described "alarm calls" of *L. curvirostra* that were "like the Scottish tooping" and were given by males and females at nests. His audiospectrogram shows no difference between the two sexes, but clear structural differences are evident between these and toops of the the Scottish form. Given the behavioral correlates he described, his "alarm calls" of *curvirostra* belong in the same category as toops or excitement calls. Excitement calls of both European forms were spectrographically similar to those of Type 2 North American crossbills in that all had a main element at slightly over 2 kHz with closely overlying frequency bands.

Chitter calls. Crossbills in interacting and foraging flocks were often heard giving a variety of soft, high-pitched chip notes, subsumed here under the name *chitter*. These differed from flight calls in that they were always given with a low amplitude and never when flocks were in full flight. Wild birds would often give chitter when fluttering to lower perches as they approached decoys, and never appeared to signal aggression. Captives would give chitter when supplied with fresh food, water, or green branches. Chitter calls were highly variable in structure. Direction of frequency modulation was either upward or alternated (Fig. 9A) or downward, as are flight calls (Figs. 9B and 9C). Chitter calls sometimes resembled flight calls, but the two categories of calls were often given in the same series without gradation (Fig. 9C). Nethersole-Thompson (1975:122)

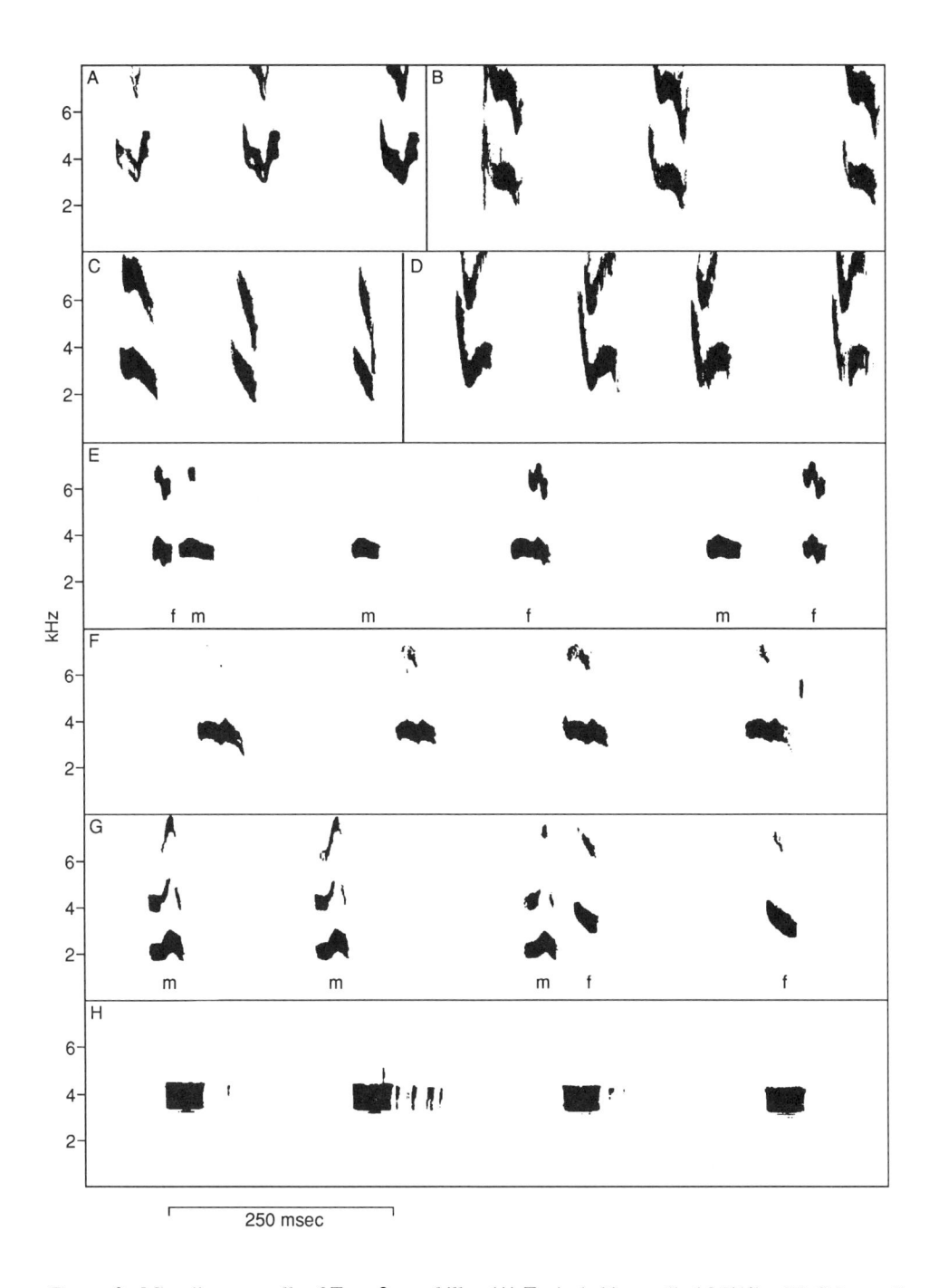

Figure. 9. Miscellaneous calls of Type 2 crossbills. (A) Typical chitter calls (aM419). (B) Chitter calls by jM737 that are structurally similar to flight calls, but with energy concentrated in upper harmonic bands. (C) Series of calls by aF541 changing from flight calls (note 1) to chitter calls (notes 2 and 3). (D) Threat calls (aM554). (E) Calls associated with courtship feeding in pair aM189/aF190. (F) Calls given prior to copulation by aM189. (G) Calls at the nest of pair aM189/aF190. (H) Nest-begging calls of aF343.

described similar calls given in the same contexts in Scottish crossbills as "a subdued twittering *wik-wik*" and in common crossbills as "twittering *peep-peeps*."

Threat calls. In hostile interactions, crossbills often combined excitement calls with a structurally distinct call termed the *threat call*. Threat calls were always given in rapid series, were relatively high in pitch, and showed strong rates of upward and downward frequency modulation (Fig. 9D). The sexes were similar in their use of this call. Although threat calls generally came after a bout of excitement calls, the two categories of calls never graded together. Threat calls were associated with increased bodily contact, such as by birds grappling with their legs or by combatants facing one another with open bills (illustrated by Tordoff 1954a:352). Threat calls were never given by birds in the hand. Little is known regarding individual variation in threat calls.

Calls during courtship feeding and copulation. Courtship feeding is an important behavioral element in the establishment of the pair bond in finches (Hinde 1955, Newton 1973). Calls (Fig. 9E) were given when food was given to the female by the male or when bills touched and no food was transferred. Both birds gave calls, but notes were inconsistent in structure and differed both within and between courtship feeding bouts. The calls bore some resemblance to flight calls but were lower in amplitude. Calls of females were often more like chitter in that they emphasized upper harmonic frequencies and were higher in pitch. Copulation was not observed in captive crossbills, but one wild male (aM189) gave calls similar to flight calls before mounting the female (Fig. 9F).

Calls at nests. When approaching nesting females after a foraging bout, males perched in nearby trees for up to one or two minutes giving excitement calls (Fig. 9G) identical to those given in other circumstances (see above). Females responded by begging from the nest with structurally different calls given in long repetitious series (Figs. 9G and 9H). Females were also observed giving begging calls toward males when both were away from nests. Structures of female begging calls differed dramatically among individuals. Some (e.g., Fig. 9G) were simple high-pitched and downward-modulated notes, whereas others were nonmodulated and high pitched, modulated upward in frequency, or had overlying harmonic bands. Females also gave typical excitement calls when their nests were disturbed, a behavior noted in other nesting crossbills (Nethersole-Thompson 1975).

CALLS OF JUVENILES

Calls of nestlings. Developing crossbills undergo a consistent series of stages of begging calls. Because of the very low amplitude of early calls, most observations were obtained by holding a microphone a few inches away from nests in captivity. When less than 6 days old, nestlings gave almost inaudible peeps (Fig. 10A). Considerable variation in pitch of these notes was evident, and it is possible that siblings differed consistently. After 6 days, calls became louder, higher in pitch, and longer in duration (Fig. 10B). After about 8 days, calling became long-duration and nearly tonal *seee* notes (Fig. 10C).

Calls of fledglings. From an age of about 12 days until several weeks after leaving the nest, juvenile crossbills gave characteristic *chitoo* begging calls (Fig. 10D). These calls were loud and conspicuous, and could be heard as juveniles followed their parents through

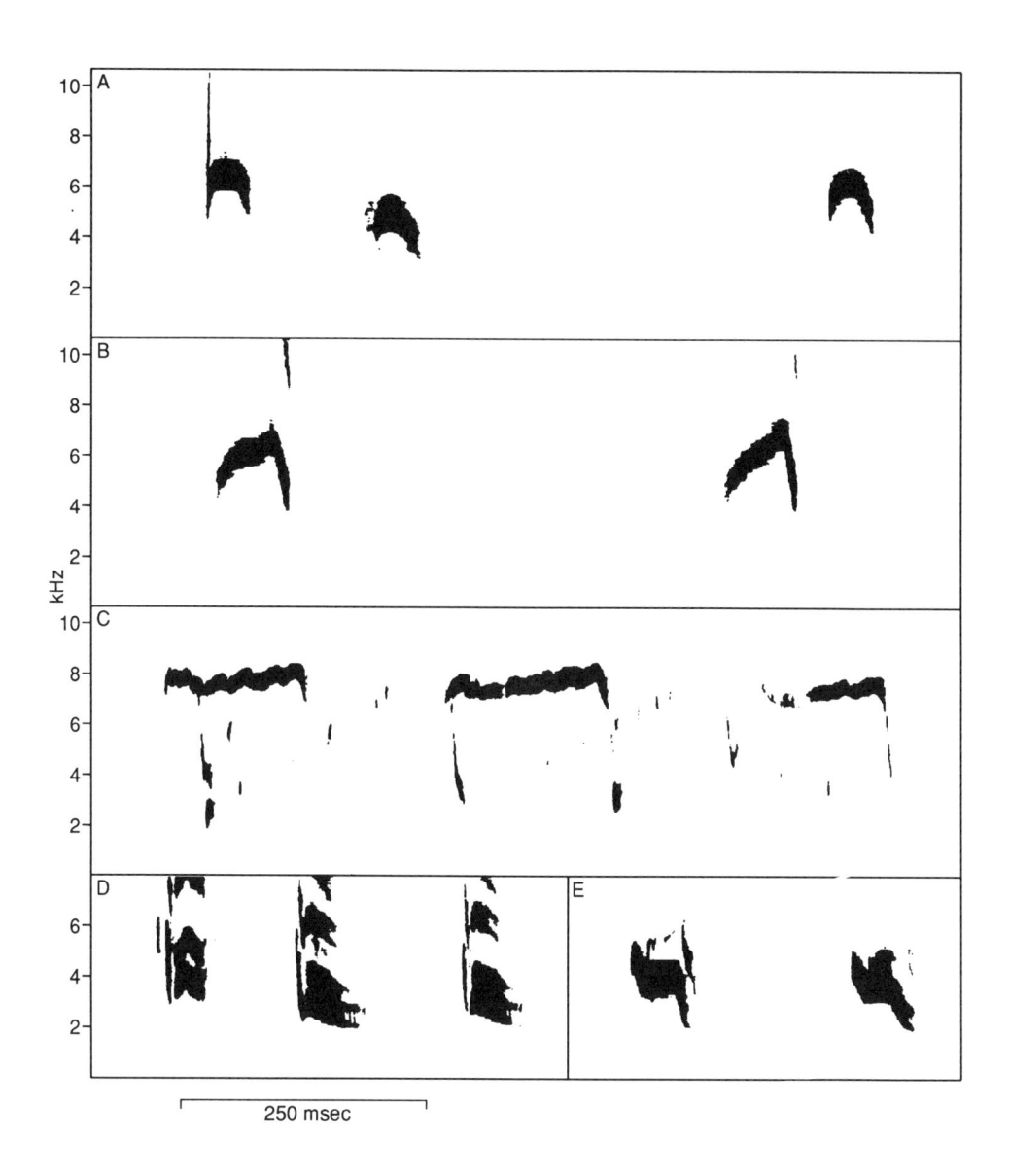

Figure 10. Ontogeny of calls in juvenile Type 2 crossbills. (A) Calls of 2-day-old nestlings (presumably two individuals). (B) Calls of a single 6-day-old nestling. (C) "Seee" calls of a single nestling between 7 and 9 days old. (D) "Chit-too" begging calls of jF521 (post-fledging). (E) Early flight calls of jF518.

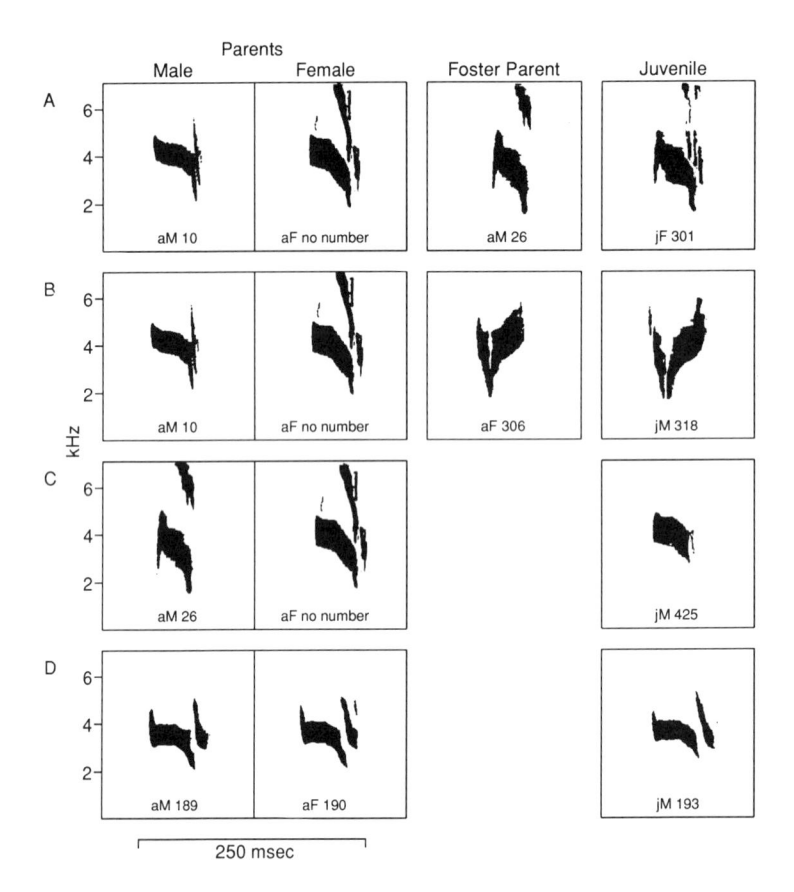

34   *University of California Publications in Zoology*

Figure 11. Development of flight call structures in juvenile crossbills of known parentage. (A) Foster rearing of jF301, daughter of Type 2 parents (aM10 and aF "no number"), with a Type 1 foster father [aM26]. (B) Foster rearing of jM318, from the subsequent brood by the same parents as jF301, with a foster parent (aF306) with structurally distinctive calls. (C) Product (jM425) of a cross between a Type 1 male (aM26) and a Type 2 female (aF "no number") in captivity. (D) Juvenile (jM193) resulting from a natural nesting of aM189 and aF190.

the forest. Chitoo calls were always comprised of two distinctive note types, with a beginning "chit" followed by from one to four downward-modulated "too" notes. Chitoos were given rapidly and monotonously in long series by juveniles, often while quivering their wings and facing adults. Nethersole-Thompson (1975) described similar calls in Scottish crossbills; his use of the term *chitoo* is followed here. Chitoo begging calls were heard given by juveniles also heard giving flight calls like those of adults on the same days. Flight calls in very young birds were often inconsistent in note structure, unlike those by adults. Early flight calls often resembled begging calls in that they were given with a "chittoo" cadence similar to begging calls (Fig. 10E).

Vocal learning. Juvenile crossbills learn flight call structures from social tutors (Fig. 11); therefore, these structures are not innate. In two instances in which experimentally tutored birds diverged from the call structures of their genetic parents (Figs. 11A and 11B), the juveniles were fed by their foster parents. Juveniles that began tutoring at more than one month of age and were independently foraging did not match the flight calls of foster parents, suggesting that learning occurs at an early age. That vocalizations have a learned component in crossbills is consistent with surveys of other carduelines in which vocal learning and imitation are prevalent (Mundinger 1979, Remsen et al. 1982, Kroodsma and Baylis 1982).

GROUPS BASED ON VOCALIZATIONS

Studies of avian vocalizations have often been instrumental in clarifying taxonomic relationships near the species level (e.g. N. K. Johnson 1963, 1980; Löhrl 1963; Payne 1973; Lanyon 1978; Ratti 1979; Nuechterlein 1981; Wells 1982; Pratt 1989; Pitocchelli 1990; Johnson and Jones 1990). Previous work by Nethersole-Thompson (1975) and Groth (1988) showed that different forms of crossbills living in the same areas have distinguishable vocalizations. This suggests that the study of vocalizations in crossbills, as in some other birds (e.g., White 1789; Stein 1958, 1963; Rowley 1967; Traylor 1979; Payne 1973, 1982), may reveal the existence of sympatric sibling species. This hypothesis would be refuted by finding that vocalizations vary continuously or consistently with geography (e.g., Goldstein 1978, N. K. Johnson 1980, Adkisson 1981, Kroodsma 1981) or show no differences over thousands of kilometers (e.g., E. H. Miller 1986).

DESCRIPTIONS OF FLIGHT CALL TYPES

All red crossbills recorded gave single-note flight calls with rapid frequency modulation. Most flight calls had the peak sound energy at 3.5–4 kHz and ranged from 40 to 80 msec. Beyond these general similarities, the shape and structure of flight calls varied greatly among individuals. Most variation involved the direction of frequency modulation over the duration of the note. Depending on the individual, audiospectrograms showed flight calls as either one continuous element or several elements separated slightly in frequency, time, or both.

Call notes fell into 8 distinctive "types" (Fig. 12). These were different enough for classification using audiospectrograms, and with some familiarity humans can detect audible differences. This classification scheme is used as a working hypothesis. Because individuals gave calls of only one type, crossbills are discussed herein with reference to type numbers (e.g., "Type 5 male," etc.), which are not linked to existing subspecies names.

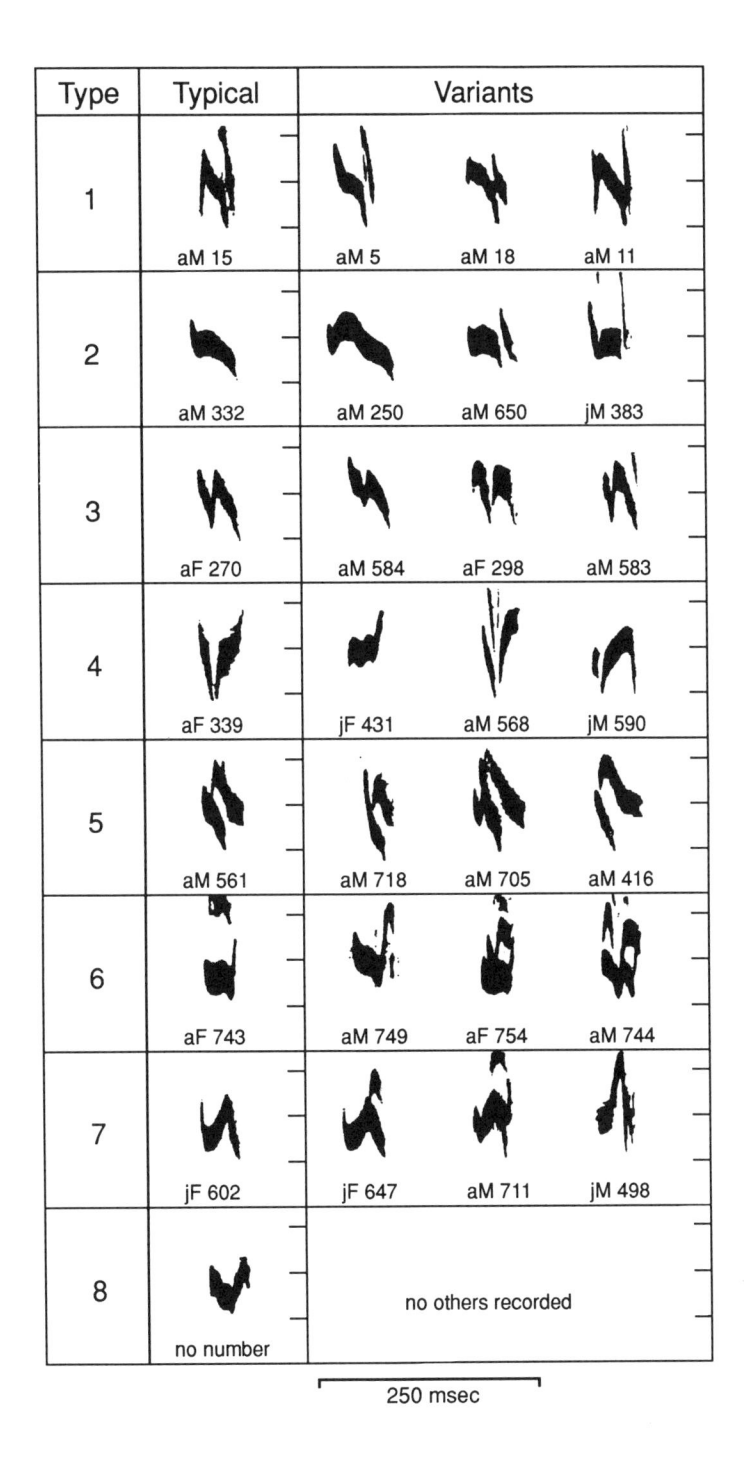

Figure 12. Representative flight calls of 8 call types of North American red crossbills. One "typical" call is shown in the left column to represent the modal call note shape for each group. Three "variant" calls are shown in the right column to show the extremes of variation within each group. Marks on the vertical scale for each row represent 2, 4, and 6 kHz.

Type 1 flight calls were dominated by a single component of rapidly dropping frequency, and almost all were initiated by an upward-modulated component of much shorter duration. Initial components were lacking in only a few individuals such as aM5, and were preceded by a rapid downward component in rare individuals such as aM11. Most birds giving Type 1 flight calls produced secondary elements immediately following the main component of the note. These ending elements were never as prominent as the secondary elements in Type 5 birds (see below).

Type 2 flight calls were characterized by a relatively gradual drop in frequency over the duration of the note. Calls of different individuals varied in duration, and although the frequency was modulated downward overall, the direction and rate of modulation at different time sections along notes varied extensively. It was common for Type 2 flight calls to have a rapid fall in frequency for the first few milliseconds and then either a slight rise or leveling before a final prolonged fall. Like Type 1 flight calls, some Type 2 calls contained ending elements separated from the main portions of the notes, as exemplified by aM650, but many had no traces of such an element.

Type 3 flight calls consisted of two waves of downward frequency modulation connected by an intervening short span of upward modulation. In most, the first downward segment began at a higher frequency than the second, with the second downward segment terminating at the lowest frequency of the note. Considerable individual variation was observed in the frequency spans of the two downward waves. Many individuals had upward modulated elements at the beginning of their call notes similar to those of Type 1 calls (as in aF298), and a minority (e.g., aM583) had downward-modulated ending components less prominent than those of Type 1 and Type 2 flight calls.

Type 4 flight calls concentrated most sound energy in an upward-modulated component. The majority had less prominent initial components modulated downward in frequency, giving the notes a *V* shape, but other birds (e.g., jF431) produced only the upward-rising element. Unlike the preceding three call types, Type 4 calls were never followed by downward-modulated ending components separated from the main portion of the notes; however, one individual (jM590) had a rapidly downward-modulated wave attached to the main upward element.

Type 5 flight calls had two elements dropping in frequency but different in frequency domains. The lower elements were generally simpler and showed less individual variation than the upper elements. Upper elements usually rose sharply before modulating downward, and the rate of downward modulation slowed near the terminus in many individuals. That the two elements were modulated differently over the same span of time may be evidence that separate halves of the syrinx operate simultaneously (Greenwalt 1968, D. B. Miller 1977) to produce Type 5 flight calls.

Type 6 flight calls had main elements nearly tonal or slightly downward-modulated in frequency, and all had an abrupt terminal rise in frequency. Calls of some individuals (e.g., aF754) contained separate *V*-shaped waves of higher absolute frequency than the lower elements, with the apexes of the *V*s overlapped by the main lower portion of the notes. Like Type 5 flight calls, the presence of two differentially modulated elements over the same time span may be evidence for simultaneous operation of separate halves of the syrynx.

Type 7 flight calls were like Type 3 in having an initial fall, middle rise, and final fall in frequency; Type 7 calls, however, had a more prolonged middle rising portion. The duration and frequency span of the initial downward portions of the notes were generally much shorter than for Type 3 calls. Some Type 7 calls appeared like inverted Vs (e.g., jM498). Half of the Type 7 calls had higher frequency elements similar to reduced versions of the upper elements of Type 5 flight calls (e.g., jF647), whereas others (e.g., jF602) did not.

Type 8. Flight calls of red crossbills from Newfoundland are represented by recordings of a single bird (Fig. 12 and Appendix C), and the short segment of recordings available allowed production of audiospectrograms of four virtually identical call notes. The calls do not match any others recorded from continental North America and were considered a distinctive type. The "Type 8" notes appeared most similar to Type 6 in that they were first modulated down and then more rapidly up in frequency, but the upward portion was stronger in the calls from the Newfoundland bird. The calls were V-shaped like Type 4 flight calls, but differed in having the initial downward section prominent and more slowly modulated.

CORRESPONDENCE BETWEEN FLIGHT CALLS AND OTHER CALLS

If birds with similar flight calls from geographically distant sites differ in other categories of calls, then flight call similarity could be explained as the result of chance convergence; alternatively, if birds match in the structure of other calls, it would support divisions based on flight calls.

Alarm calls. Alarm calls differed consistently among groups based on flight calls (Fig. 13, left). Type 1 alarm calls were spectrographically simple in that they were short–duration unmodulated whistles at $\pm$ 2.5 kHz. Type 2 alarm calls had more emphasis on upper harmonic bands, and the main elements were modulated first upward then downward in frequency. Type 3 alarm calls were structurally complex, with two simultaneous elements of frequency modulation, both with strong harmonics; upper elements were longer in duration and at approximately 3.5 kHz, whereas lower elements were around 2.5 kHz and shorter in duration. Type 4 alarm calls were similar to those of Type 1 in that they were generally unmodulated whistles; however, Type 4 alarm calls were consistently longer in duration and many wavered in frequency. Type 5 alarm calls were among the shortest in duration and contained strong harmonics, and like those of Type 3 consisted of two simultaneous elements of frequency modulation; standard "wide-band" spectrograms (as in Fig. 13) do not clearly reveal this structure, whereas "narrow-band" graphs (not illustrated) show the two elements separated by 0.3–0.5 kHz. Unlike those of other types, Type 6 alarm calls showed an overall drop in frequency, and were most similar to those of Type 5 in that they were short in duration and had simultaneous elements separated by 0.3–0.5 kHz (also not clearly visible with wide-band spectrograms);

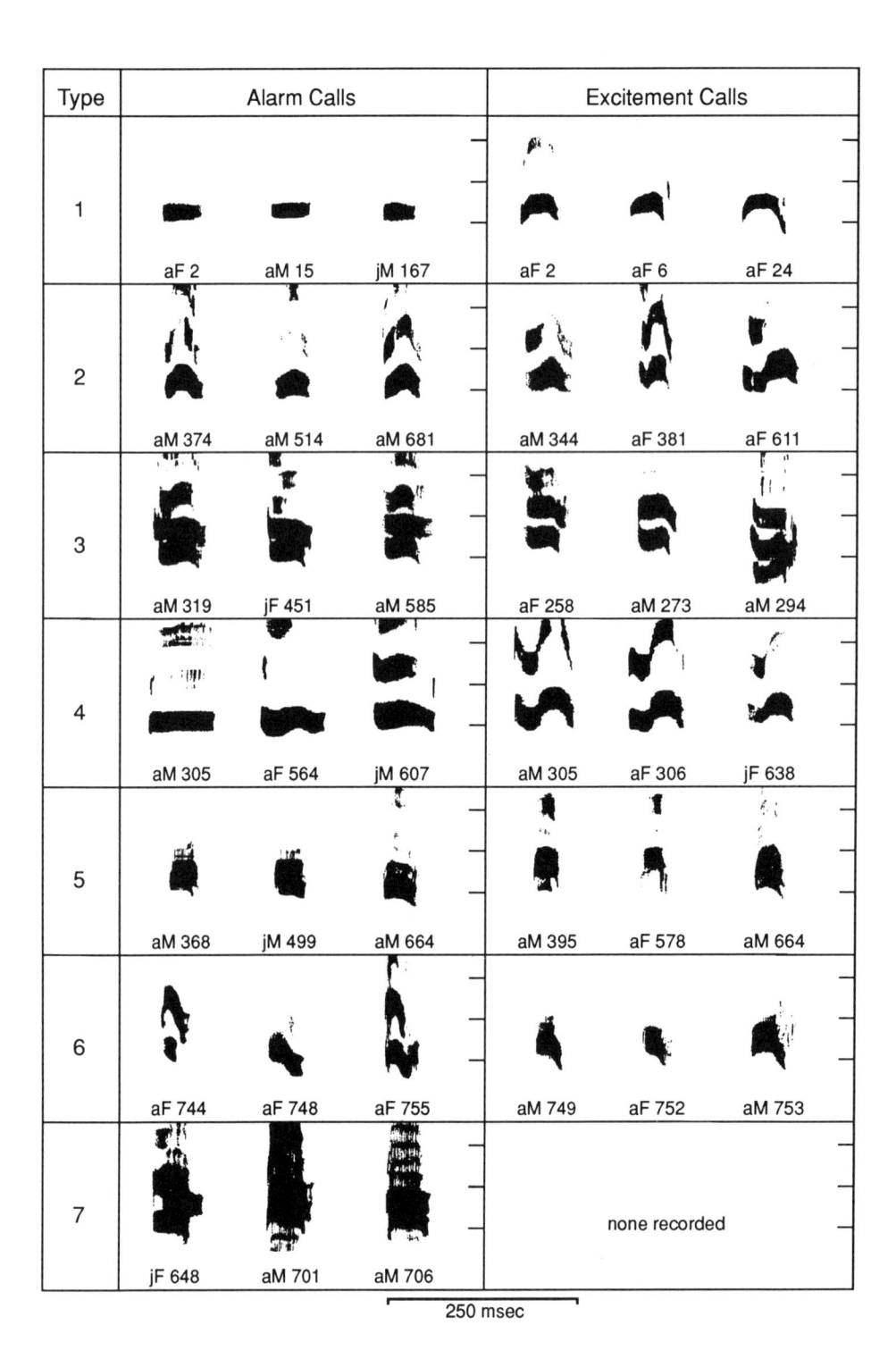

Figure 13. Comparison of alarm calls (left) and excitement calls (right) in 7 groups of red crossbills divided on the basis of flight call structure. A single call from each of three individuals is shown for each category of call within each group. Marks on the vertical scale for each row represent 2, 4, and 6 kHz.

harmonics of the top band were often emphasized (e.g., aF744). Type 7 alarm calls were like long-duration versions of those of Type 5, and harmonics were strongly emphasized.

Excitement calls. This call was recorded for all call types except Type 7 and Type 8 (Fig. 13, right). Type 1 excitement calls were simple in structure and chevron-shaped, resembling high-frequency versions of Type 2 alarm calls. Type 2 excitement calls were more complex, usually dropping in frequency before rising and again falling; upper "harmonic" bands were emphasized in the first half of the notes, and lower bands were stronger in the final rise-and-fall portions. Type 3 birds gave excitement calls consisting of two separate elements similar to their alarm calls, except that their excitement calls averaged higher in pitch. One male (aM294) gave unusual excitement calls that were far lower in pitch than those of other Type 3 birds. Type 4 excitement calls were virtually identical to those of Type 2, but it appeared that Type 4 excitement calls averaged longer in duration. The excitement calls of Type 5 birds were the shortest in duration and can be described as high-pitched versions of Type 5 alarm calls; "narrow-band" spectrograms reveal a layering of separate elements differing in frequency and direction of modulation, as in alarm calls. The same close layering of elements was found among Type 6 excitement calls, which were also were similar to Type 6 alarm calls except higher in pitch.

FLIGHT CALL VARIATION WITHIN GEOGRAPHIC REGIONS

One of the major questions asked in this study is whether vocally defined forms of crossbills live and breed in the same areas. It is also critical to ask to what extent vocally different kinds of birds interact socially. Below, a description of the call type composition within geographic regions (see Fig. 2) is presented, including audiospectrograms of flight calls of all birds of known morphology (Figs. 14 to 21). Each figure in this section is divided into "series" by locality (generally presented in a north–south sequence) and date, and flight calls notes are labeled with the age, sex, and individual identification number of each bird. A listing of individuals and the call types to which they were assigned is contained in Appendix E. This section contains a synopsis of all that is currently known regarding the geography of variation among North American red crossbills.

Northwestern region. Six of the eight known call types are represented by birds recorded in this region (Fig. 14 and Appendix C); only Type 6 and Type 8 have not been recorded. The northernmost recordings obtained for this study were taken at two sites near Prince George, British Columbia, in early August 1987. One flock of four Type 7 juveniles was recorded, from which one (jF602) was collected near the Willow River, where both Type 3 and Type 4 birds were also recorded. A few miles east, along the Bowron River, a few juvenile Type 3 and Type 4 were recorded and collected, but no mixed flocks were observed. The most vocally diverse assemblage of crossbills found to date was encountered on the Thompson Plateau east of the Fraser River in central British Columbia, where six call types were recorded. Two Type 1 individuals (birds 635 and 636) were taken, providing the only specimens of this type outside of the Appalachians (aF636 had slight follicular enlargement but no incubation patch, and jM635 had testes about half breeding size). Type 2 birds were common at this locality. All of the Type 2

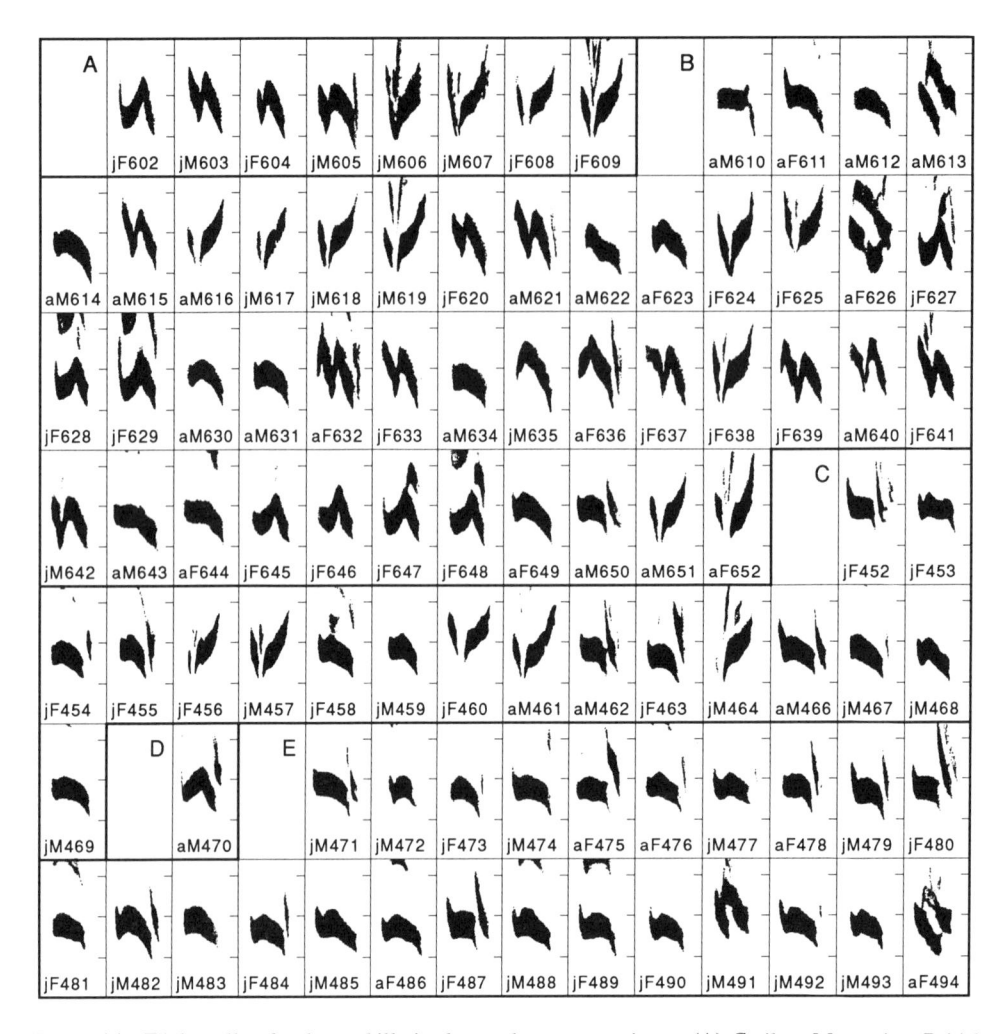

Figure 14. Flight calls of red crossbills in the northwestern region. (A) Cariboo Mountains, British Columbia, 1–3 August 1987. (B) Thompson Plateau, British Columbia, 4–10 August 1987. (C) Northeastern Washington, 29–31 July 1986. (D) Northern Idaho, 2 August 1986. (E) Sawtooth Mountains, Idaho, 4–6 August 1986. For each audiospectrogram, the frequency markers on the vertical axis are at 2, 4, and 6 kHz. The width of the box for each note is equivalent to 84 msec. Individuals are labeled by age (j=juvenile [age classes 0 to 6, Table 1], a=adult [age classes 7 to 9, Table 1]); sex (M=male, F=female, O=sex undetermined), followed by an individual identification number. Appendix A contains specific locality data, and Appendix E contains call type designations.

males taken were fully plumaged adults with enlarged testes. The three Type 2 females had some enlargement of the ovaries, and one (aF611) had an edematous incubation patch. The other two females had featherless but less vascularized bellies. Type 3 birds were either juveniles or breeding (or recently post-breeding) adults. Both juvenile and adult Type 4 birds were found; aF652 had a nonvascular but conspicuous incubation patch with slight ovarian enlargement, and aM651 (probably her mate) had testes of breeding size. Representing Type 5, aM613 was near breeding condition and aF626 had just laid, having three collapsed follicles; another flock of Type 5 birds had adults feeding begging juveniles. All Type 7 birds encountered were collected, all in two flocks (birds 627–629 and 645–648) consisting only of juvenile females in fully streaked plumage and with undeveloped ovaries. Crossbills encountered at this locality were traveling as lone individuals or small flocks, but no groups were of mixed call type.

At Sheep Creek in northeastern Washington in July 1986, both Type 2 and Type 4 birds were recorded, most of which were juveniles traveling in small flocks. The two call types were closely associated because of the density of the population, but flocks usually contained a single type. Types 1, 2, 3, 4, and 5 were recorded in western Washington in the late 1980s (Appendix C). The Type 1 calls from Washington were accompanied by alarm calls and excitement calls showing precise matching to those of southern Appalachian Type 1 birds described below.

Few crossbills were encountered in the panhandle of northern Idaho in summer of 1986, but a male–female pair of Type 2 birds was recorded (not shown), and a lone Type 7 male (aM470) was collected which had only partially enlarged testes. In the Sawtooth Mountains of Idaho in August of 1986, most crossbills were Type 2 juveniles in small flocks, suggesting breeding nearby in preceding months. Type 5 birds were less common at this locality, and two were collected (jM491 and aF494; the female had a post-breeding incubation patch and slightly enlarged ovary). A flock of six Type 4 birds (not shown) stopped in the trees near the camp and was recorded.

Oregon. Five different call types have been recorded in Oregon (Fig. 15 and Appendix C). Along the Oregon coast only Type 3 and Type 4 crossbills have been recorded. In July of 1986 in the coastal forests north of Florence both call types were abundant, and both contained young juveniles and members in breeding condition. Crossbills were not common at the same site in 1988, and of the three crossbills collected, the female of a Type 4 pair (aM698 and aF699) was laying, but a lone Type 3 female (aF700) was not in breeding condition. A small number of nonbreeding Type 3 and Type 4 birds were recorded on the western slope of the Cascades in June of 1985. The call type distribution in eastern Oregon was different from that on the coastal slopes. On the east slope of the Cascades, only Type 2, Type 5, and Type 7 crossbills have been recorded. Only Type 2 birds were recorded in the area around Chiloquin from 6–7 June 1985 and mid-April 1988, but at higher elevations both Type 2 and Type 5 birds were common and breeding in August 1986 and July 1988. Breeding Type 7 birds were also found in these mixed populations of the eastern Cascades, but they were not common. A few Type 2 crossbills were recorded in the northern extension of the Warner Mountains in July 1988. Other observers have recorded Type 2 crossbills in eastern Oregon (Appendix C).

Northern and coastal California. Four call types have been recorded in this region (Fig. 16 and Appendix C). In late 1984 and early 1985, the San Francisco Bay area experienced a large influx of thousands of crossbills of at least four different call types, with some flocks containing up to several hundred birds. Type 2, Type 3, and Type 4

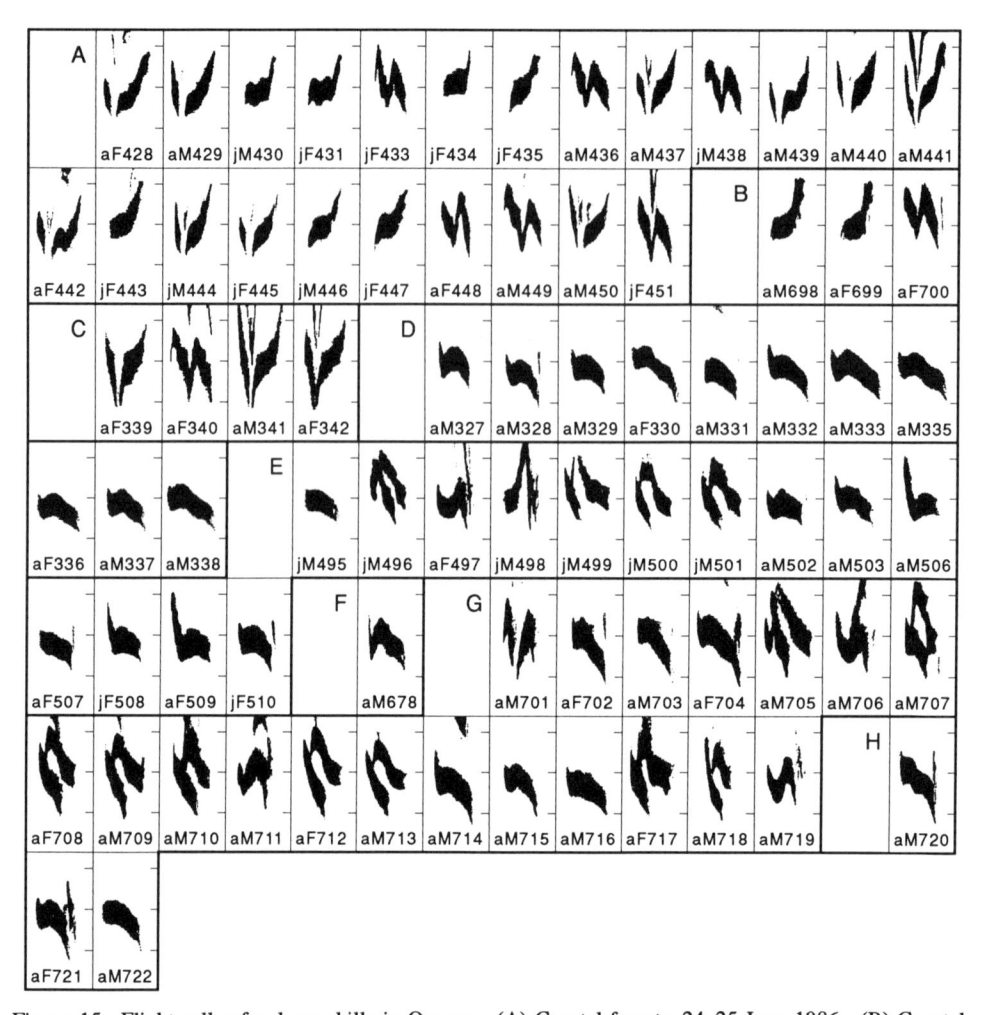

Figure 15. Flight calls of red crossbills in Oregon. (A) Coastal forests, 24–25 June 1986. (B) Coastal forests, 22–23 July 1988. (C) Western Cascade Mountains, 19 June 1985. (D) Eastern Cascade Mountains, 6–7 June 1985. (E) Eastern Cascade Mountains, 11–13 August 1986. (F) Eastern Cascade Mountains, 18 April 1988. (G) Eastern Cascade Mountains, 24–26 July 1988. (H) Warner Mountains, 27 July 1988. Scale and legend as in Fig. 14.

birds were most common. Type 5 was recorded less commonly, and no birds giving calls of this type were collected during this invasion. Flocks generally contained a single call type; as an example, birds 217 to 227 comprised an entire flock. Pine siskins (*Carduelis pinus*) were also abundant, and they often flocked with the crossbills. There is no evidence that any of the invading crossbills nested in the region. In the mountains near Hyampom, California, in late May and early June 1985, both Type 3 and Type 4 were recorded. These may have been remnants from the large invasion of the previous winter. Twice flocks of 20–30 birds containing both call types were noted, and the series birds is

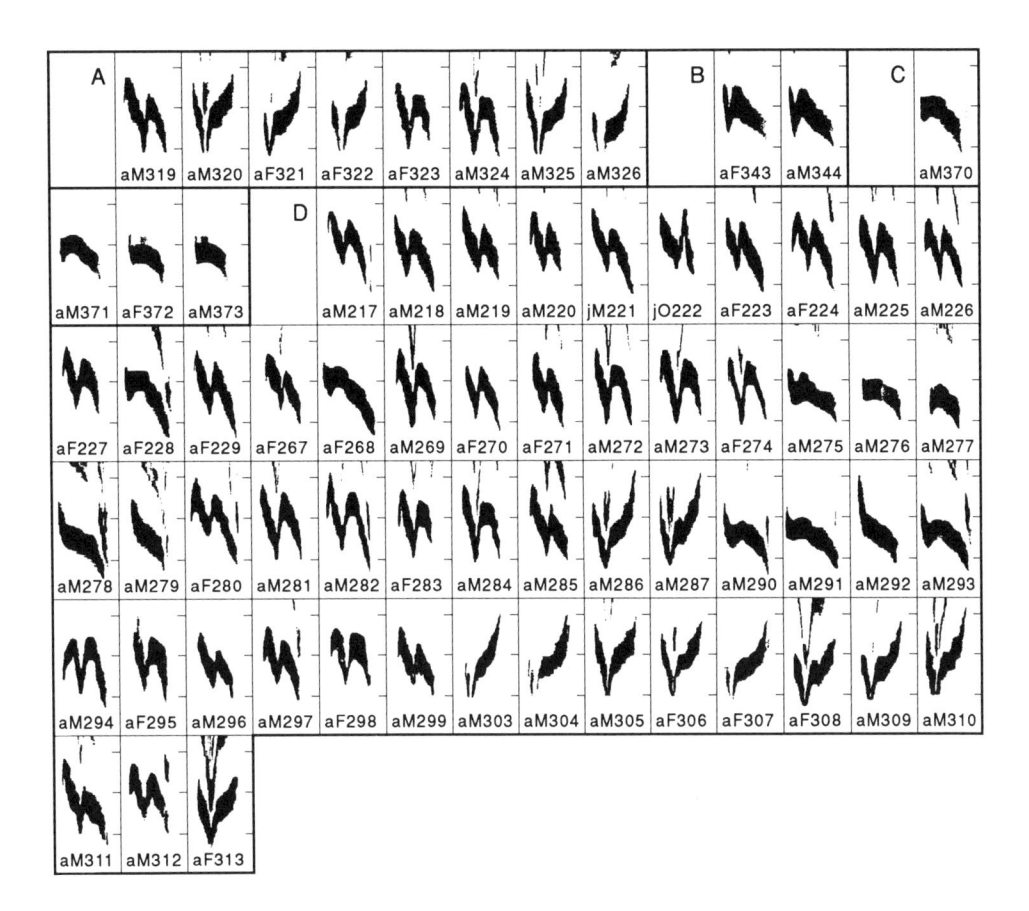

Figure 16. Flight calls of red crossbills in northern and coastal California. (A) Northern Coast Range, 31 May 1985. (B) Northern Coast Range, 25 June 1985. (C) Northern Coast Range, 22–24 August 1984. (D) San Francisco Bay area, 25 August 1984–13 March 1985. Scale and legend as in Fig. 14.

from a single flock. All were adults, and none appeared to be breeding. The following month, in the Coast Range overlooking the Central Valley in Glenn County, a breeding population with a single vocal type (Type 2) was recorded. The two birds collected on 25 June 1985 (birds 343 and 344) were a mated pair taken from a nest.

The Sierra Nevada, California. Type 2 was the most frequently recorded vocal type in the Sierra Nevada (Fig. 17 and Appendix C). Coincident with the large 1984–85 invasion of crossbills on the California coast, crossbills were abundant at Sagehen Creek in the Sierra Nevada in October 1984; however, almost all Sierra Nevadan crossbills gave Type 2 calls, with the exception of a Type 3 pair (birds 258 and 259). All specimens in this series were nonbreeding. Type 2 calls were commonly heard and recorded in the Sierra Nevada in subsequent years. An exceptional flock of approximately 20 Type 5 birds was recorded at Sagehen Creek on 16 February 1987, with flocks of 40–60 Type 2 birds

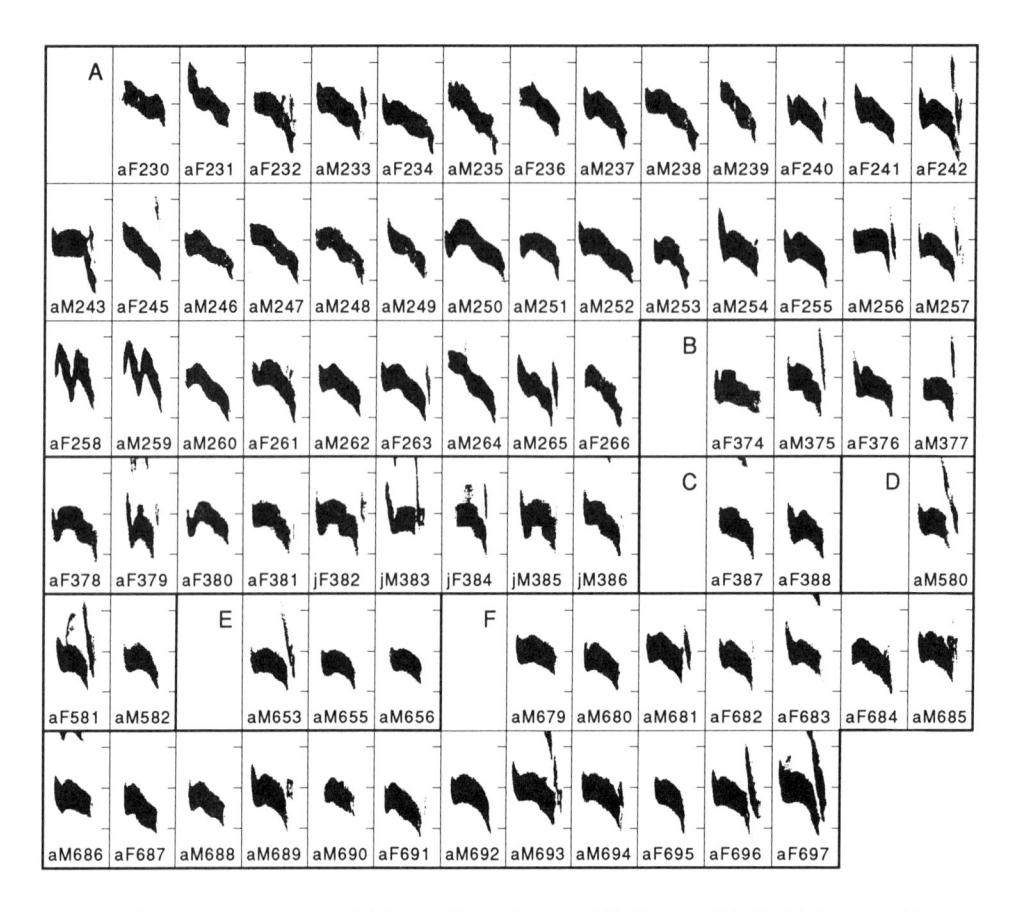

Figure 17. Flight calls of red crossbills in the Sierra Nevada of California. (A) 13–14 October 1984. (B) 26–27 October 1985. (C) 3 May 1986. (D) 11 June 1987. (E) 31 August 1987. (F) 29 June–1 July 1988. Scale and legend as in Fig. 14.

also recorded on the same day, but mixed flocks were not observed. A single Type 7 individual in an otherwise homogeneous flock of Type 2 birds was also recorded that day, providing the southernmost recording locality for this type. The breeding crossbills recorded in the summer of 1988 at Sagehen Creek gave Type 2 calls exclusively. A single Type 5 bird (not shown) was recorded in the northern Sierra Nevada where Type 2 birds were feeding their young in August 1987. Type 2 crossbills have been recorded around Yosemite in the summers of 1981 and 1984 (Appendix C).

Arizona and southern Utah. Five different call types have been recorded in Arizona (Fig. 18 and Appendix C), yet most of the recordings from this state have been of Type 2 birds. On the Kaibab Plateau, Arizona, in June 1986, Type 2 adults were in small groups and pairs and apparently preparing to breed. Type 2 crossbills on the Mogollon Rim in 1985 were common and nesting, but at the same locality in May 1986 only a single small flock (Type 2 adults and juveniles, all collected) was encountered. Only rare Type 2 birds

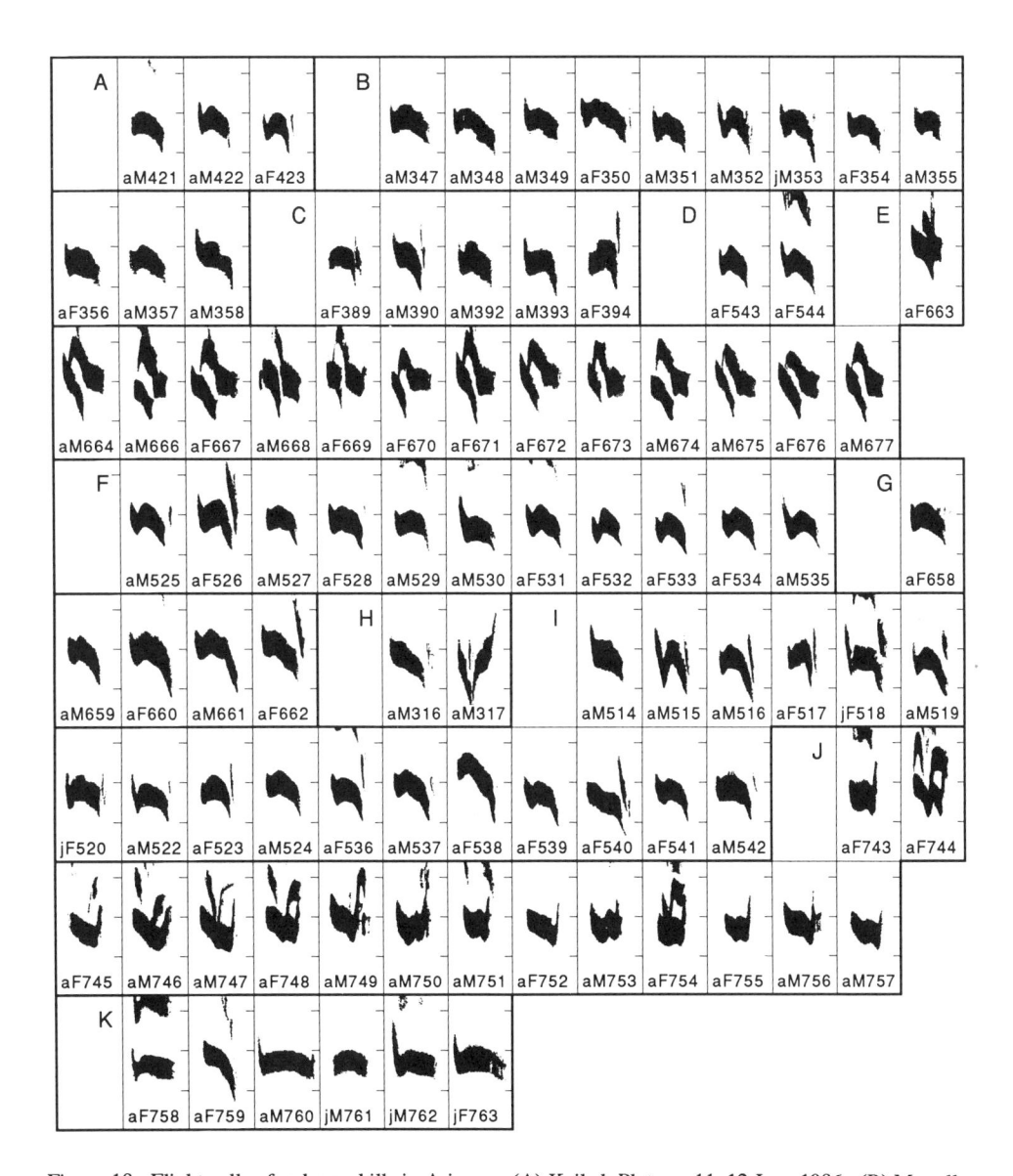

Figure 18. Flight calls of red crossbills in Arizona. (A) Kaibab Plateau, 11–12 June 1986. (B) Mogollon Rim, 27 July–1 August 1985. (C) Mogollon Rim, 28 May 1986. (D) Santa Catalina Mountains, 30 May 1987. (E) Santa Catalina Mountains, 3–4 April 1988. (F) Pinaleno Mountains, 25 May 1987. (G) Huachuca Mountains, 30 March 1988. (I) Chiricahua Mountains, 22–28 May 1987. (J) Chiricahua Mountains, 3–14 October 1988. (K) Chiricahua Mountains, 23 May 1989. Scale and legend as in Fig. 14.

were in the Santa Catalina Mountains in late May 1987, and the two females collected were not breeding. In sharp contrast, only nonbreeding Type 5 birds were found in the Santa Catalinas in early April 1988. Both Type 2 and Type 4 birds were common in the Pinaleno

Mountains in July of 1986, and a single Type 3 bird was recorded there (Appendix C, the douglas fir [*Pseudotsuga menziesei*] crop was exceptional at that time [Joe T. Marshall, pers. comm.]). It is not known whether crossbills bred there. At the same locality in May 1987, a few adult Type 2 birds were recorded and one entire flock was taken (birds 525–535), but no Type 3 or Type 4 birds were heard, and the cone crop had been depleted. The Type 4 bird recorded farthest south in this study was aM317, captured in the Chiricahua Mountains of southeastern Arizona in May 1985 with a Type 2 male (aM316). Both males had enlarged testes. Type 2 crossbills were abundant in the Chiricahua Mountains in May 1987, but none was in breeding condition (males had only partially enlarged testes); however, begging juveniles were observed being fed by adults, suggesting that some had bred there in preceding weeks. In October 1988, only Type 6 crossbills were recorded in the Chiricahuas, all of which were nonbreeding adults. Several months later at the same locality in May 1989 no Type 6 birds were heard and only a single small flock of Type 2 birds was encountered (three juveniles were seen being fed by the adult male, and aF697 had an old incubation patch). The only available recordings from Utah (Appendix C) were of a singing Type 2 male, in what was noted by George Reynard as an "apparently territorial breeding area."

Colorado and New Mexico. Types 2, 4, and 5 crossbills have been recorded in Colorado and New Mexico (Fig. 19 and Appendix C). In 1985 the high elevation area on the White River Plateau of northwestern Colorado contained only uncommon Type 5 crossbills which were in breeding condition. In the Sawatch Mountains near Leadville, Colorado, in early June 1986, both Type 5 and Type 2 were observed. Birds of the two call types generally did not flock together, but on one occasion a Type 5 male (bird 416) with a flock of 15–20 Type 2 birds broke from the flock and approached the Type 5 decoys while the remainder of the flock continued onward. Both call types had individuals at or near breeding condition. Rare nonbreeding Type 2 crossbills were found on the Uncompahgre Plateau on 3–4 August 1985. At Spring Creek Pass in the San Juan Mountains of southwestern Colorado in June 1987, Type 2, Type 4, and Type 5 birds were abundant, and juveniles of all three types were seen. The two adult male Type 5 birds had enlarged testes, and a Type 5 female (aF571) was laying. No other specimens collected here were in breeding condition, but aF567 (Type 4) had a refeathering brood patch. White-winged crossbills (*L. leucoptera*) were also breeding locally (Groth 1992).

Crossbills nesting in the Jemez Mountains of New Mexico in July 1984 gave Type 2 calls (Appendix III, C. Benkman, pers. comm.). In the Zuni Mountains of New Mexico in August 1988 a population consisting of only Type 2 birds was found. The birds were singing and beginning to nest; all females were laying and the males had testes of maximal size. The previous year, adult Type 2 birds in the Black and San Mateo Mountains of southwestern New Mexico were in small flocks; the males had half-developed testes and the females had heavy fat but no gonadal enlargement.

Southern Appalachians. Only Type 1 and Type 2 have been recorded in this region (Fig. 20 and Appendix C). Birds of both forms bred in this region in 1983 and 1984. Many of the adults examined in Montgomery County, Virginia, were in breeding condition and observations a nest and other adults feeding young juveniles were made. The post-

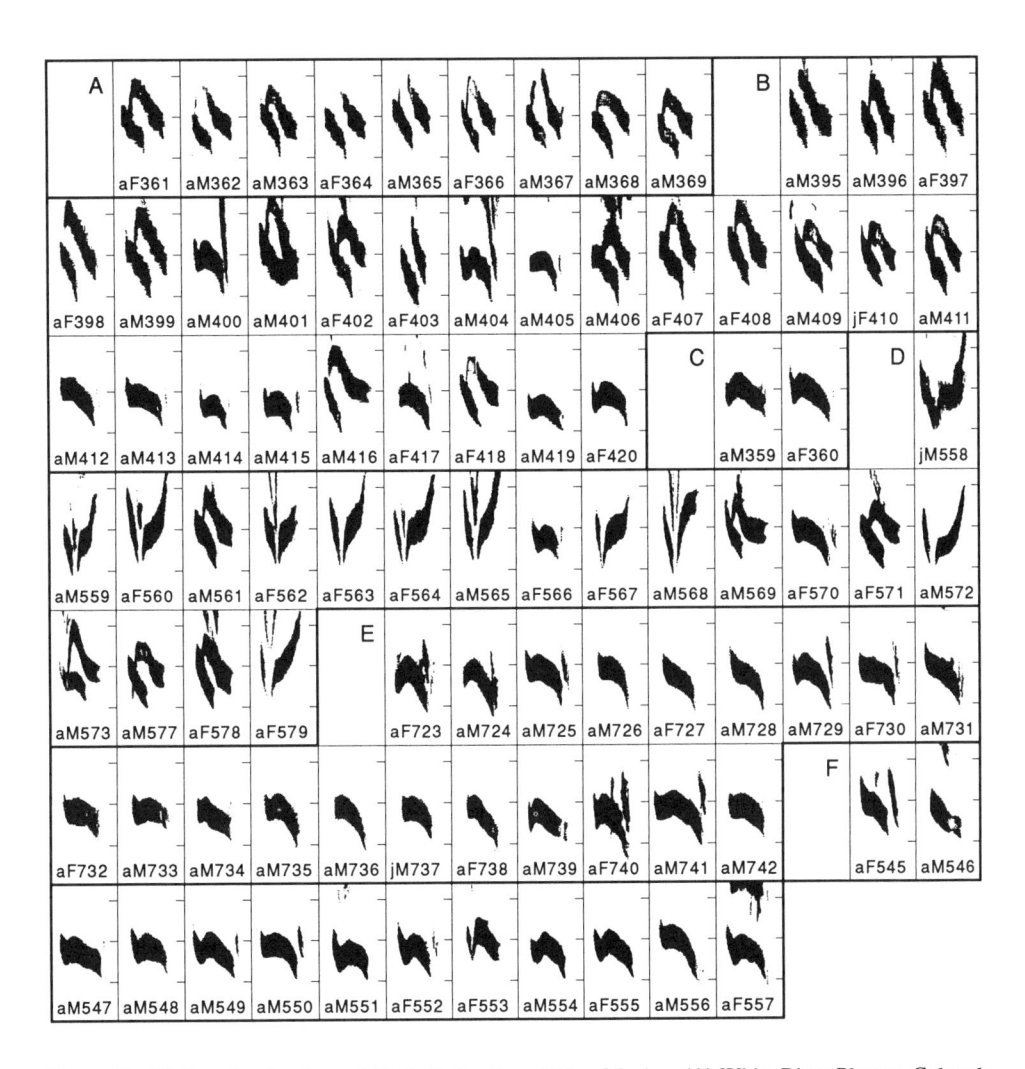

Figure 19. Flight calls of red crossbills in Colorado and New Mexico. (A) White River Plateau, Colorado, 5–7 August 1985. (B) Sawatch Mountains, Colorado, 4–8 June 1986. (C) Uncompahgre Plateau, Colorado, 3–4 August 1985. (D) San Juan Mountains, Colorado, 4–9 June 1987. (E) Zuni Mountains, New Mexico, 14–17 August 1988. (F) Black and San Mateo Mountains, New Mexico, 31 May–3 June 1987. Scale and legend as in Fig. 14.

breeding crossbills at Highlands, North Carolina in late 1983 consisted almost exclusively of Type 1 birds (jM138 was a Type 2), with some adults feeding juveniles. Crossbills recorded and collected on Roan Mountain, North Carolina, were all Type 1 birds, with the exception of jF210 which was a Type 2.

Northeastern region. Type 2, Type 3, Type 4, and Type 8 flight calls have been recorded in this region (Fig. 21 and Appendix C). Of these, only one (Type 8) has not also been recorded in other regions. The other three call types, all recorded in Hamilton

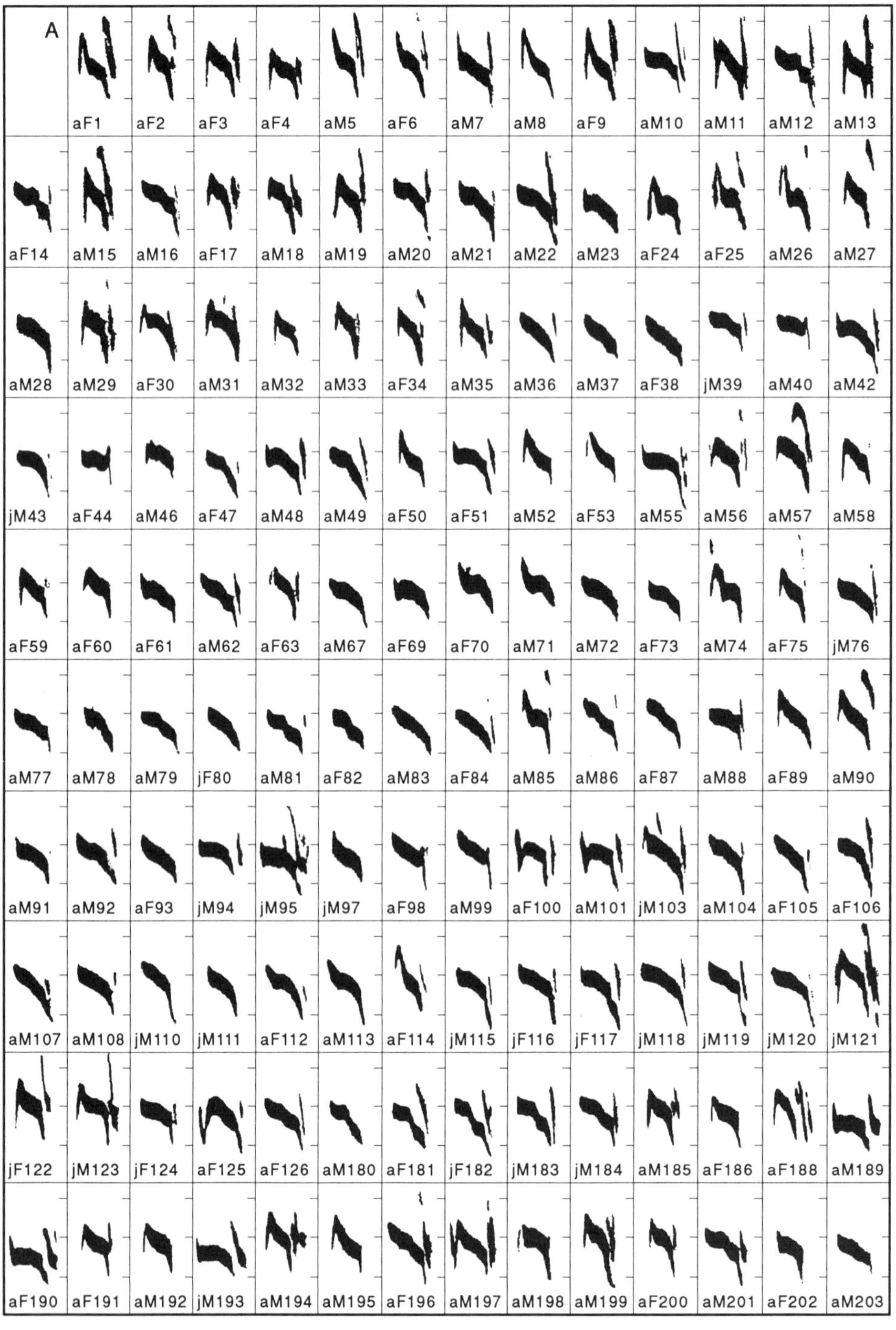
A
aF1 aF2 aF3 aF4 aM5 aF6 aM7 aM8 aF9 aM10 aM11 aM12 aM13
aF14 aM15 aM16 aF17 aM18 aM19 aM20 aM21 aM22 aM23 aF24 aF25 aM26 aM27
aM28 aM29 aF30 aM31 aM32 aM33 aF34 aM35 aM36 aM37 aF38 jM39 aM40 aM42
jM43 aF44 aM46 aF47 aM48 aM49 aF50 aF51 aM52 aF53 aM55 aM56 aM57 aM58
aF59 aF60 aF61 aM62 aF63 aM67 aF69 aF70 aM71 aM72 aF73 aM74 aF75 jM76
aM77 aM78 aM79 jF80 aM81 aF82 aM83 aF84 aM85 aM86 aF87 aM88 aF89 aM90
aM91 aM92 aF93 jM94 jM95 jM97 aF98 aM99 aF100 aM101 jM103 aM104 aF105 aF106
aM107 aM108 jM110 jM111 aF112 aM113 aF114 jM115 jF116 jF117 jM118 jM119 jM120 jM121
jF122 jM123 jF124 aF125 aF126 aM180 aF181 jF182 jM183 jM184 aM185 aF186 aF188 aM189
aF190 aF191 aM192 jM193 aM194 aM195 aF196 aM197 aM198 aM199 aF200 aM201 aF202 aM203

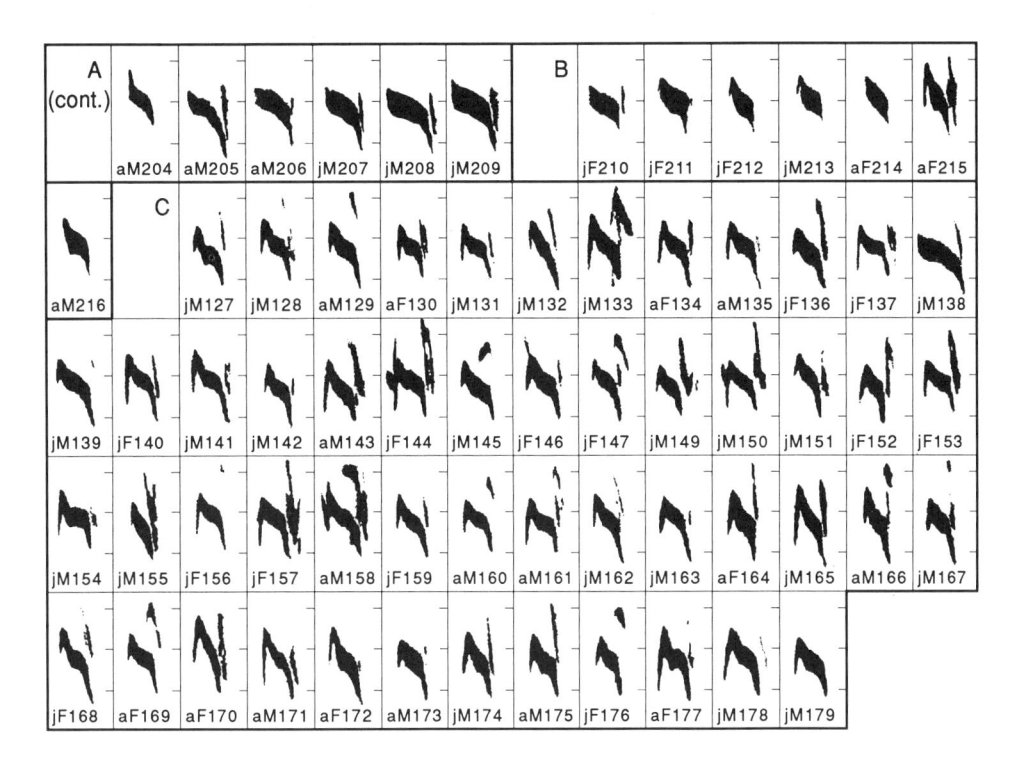

Figure 20 (with opposite page). Flight calls of red crossbills in the southeastern region. (A) Montgomery Co., Virginia, 22 February 1983–17 June 1984. (B) Roan Mountain, North Carolina, 20–21 June 1984. (C) Highlands, North Carolina, 30 October–21 November 1983. Scale and legend as in Fig. 14.

County, New York, in early 1985 were from areas in which birds were nesting (see Peterson 1985, Messineo 1985). The Type 2 birds around Albany, New York, in 1982 (Appendix III, C. Benkman pers. comm.) were breeding. Type 4 has been recorded in Nova Scotia, Maine, New York, and Ontario (Appendix III), and the Nova Scotia flock contained juveniles (C. Benkman, pers. comm.). Crossbills were uncommon in northern Michigan in July 1987, and Type 2, Type 3, and Type 4 were recorded; Type 4 and Type 2 birds were in breeding condition with adults of these types were feeding flying juveniles.The four Type 3 adults (birds 583 to 586) were near breeding condition; the males were recorded in song and had enlarged testes.

SUMMARY

North American red crossbills can be divided into a limited number of distinctive groups based on flight calls. Alarm calls and excitement calls show strong differences between groups, supporting the hypothesis for divisions based on flight calls. Individuals rarely flocked with crossbills of alien types. Despite differences on spectrograms, all flight

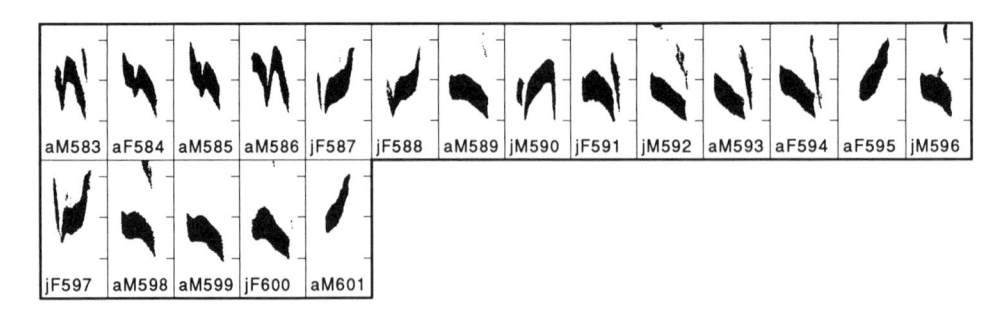

Figure 21. Flight calls of red crossbills in the northeastern region. The entire series was from the south shore of Lake Superior, 14-16 July 1987. Scale and legend as in Fig. 14.

calls, alarm calls, and excitement calls can be described by the term "chip." Considerable practice may be necessary for human observers to learn these differences and use them reliably in the field.

There are 8 different call types presently known for North America, which might be an underestimate. However, the repetitious appearance of call types at geographically extremely distant sites (see Fig. 22) suggests that not many remain undetected. With the exception of the crossbill recorded in Newfoundland, calls recorded by other observers (Appendix C) matched one or more of the call types given by the set of known birds analyzed in this study.

Call types that were found at widely separated sites did not show obvious patterns of within–group geographic vocal variation. For example, Type 2 birds in British Columbia, California, or Arizona did not have consistent differences from Type 2 birds from Michigan or the southern Appalachians. A more refined analysis of within-group variation using quantification of call note features (e.g., Sparling and Williams 1978, Groth 1988) may be feasible, but this variation would be slight when compared to the differences among call types. No features that could be used in further dividing call types into even smaller units have been identified.

Type 1 crossbills were recorded over several successive years in the southern Appalachians and appear to be relatively common there. The nearest other recording localities were over 3500 km away in Washington and British Columbia. Earlier I (Groth 1988) hypothesized that Type 1 was an Appalachian endemic, but the occurrence of vocally identical crossbills in the Pacific Northwest negates this idea. The present distributional gap is probably the result of lack of sampling and not absence of birds.

Type 2 crossbills were common and widely distributed across North America. These birds were recorded as far east as Maine and have bred in New York and Virginia. There is a fairly continuous series of recording localities proceeding west, including breeding birds in northern Michigan. The Type 2 birds at Bismarck, North Dakota, included adults with begging juveniles (R. N. Randall, pers. comm.). Type 2 birds also have a broad north-to-south distribution over the Rocky Mountain west and have bred from British Columbia to Arizona. Although Type 2 birds are the predominant crossbill in the Sierra Nevada of

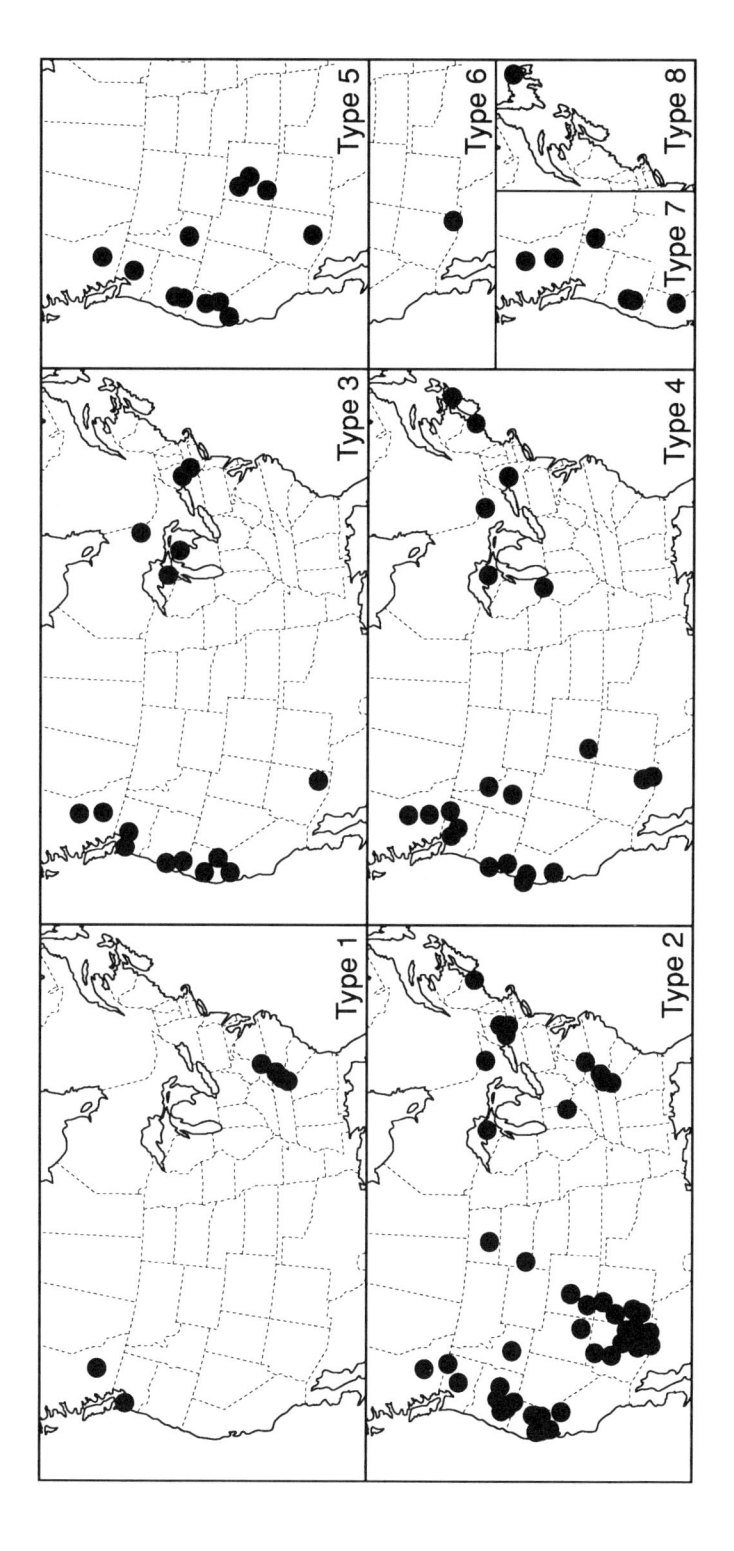

Figure 22. Summary maps of recording localities for eight call types of red crossbill. Each dot shows the position of a locality at which at least one individual of the given call type has been recorded.

California and have been recorded along the Pacific coast around San Francisco Bay, no recordings of this type have been made along the Oregon coast.

Type 3 crossbills occur over a broad area. The recordings from New York in 1985 (Appendix C) were from areas in which crossbills were breeding. The Type 3 birds collected in Michigan in July of 1987 were in breeding condition. Juveniles giving calls of this type were common in British Columbia in 1987, and adults collected there were in breeding condition. Recordings of Type 3 crossbills along the Pacific coast, many of breeding birds, are substantial. This form has been recorded as far south as the Pinaleno Mountains of southern Arizona.

Type 4 crossbills have been recorded along both the Atlantic and Pacific coasts and in many intervening areas. Recordings of Type 4 have been made as far south as the Chiricahua Mountains of Arizona and as far north as central British Columbia. This form breeds at least as far east as northern Michigan and probably all the way to the Atlantic coast. Type 4 birds are common breeders in the Pacific Northwest. The juveniles recorded in Colorado suggest breeding in the central Rocky Mountains.

Type 5 crossbills have been recorded only in the Rocky Mountain west. In Oregon the form has been recorded only on the east slope of the Cascades, not on the coast. This form apparently breeds from British Columbia south to Colorado. It is not known if birds of this vocal type breed in California, but recordings from the state exist from both summer and winter. The southernmost recordings of Type 5 birds, from the Santa Catalina Mountains of Arizona, were of nonbreeders.

Type 6 crossbills are known only from those recorded in the Chiricahua Mountains of southeast Arizona in October 1988. It is probable that these birds range most commonly in Mexico and rarely reach the United States in any numbers. Most field work in the Chiricahua Mountains and other sites in Arizona and New Mexico was with Type 2 crossbills.

Type 7 crossbills were recorded from several localities from a narrow area in the Pacific Northwest. Juveniles were collected at two localities in British Columbia. Adults, all in breeding condition, were recorded in Idaho and central Oregon. One Type 7 bird was recorded in the Sierra Nevada of California. Like Type 5, lack of recordings from eastern North America may result from a lack of sampling and not from an absence of birds.

Type 8 calls were recorded from a single individual in Newfoundland in 1981. Based on museum skin data (Bent 1912, Payne 1987, this study), Newfoundland contains a morphologically distinctive form of red crossbill; therefore, this single individual may have been a member of an endemic island form. It is not known if mainland crossbills ever reach Newfoundland, nor is it known if Type 8 birds occur on the mainland.

SOURCES OF INTRAPOPULATION VARIATION IN MORPHOLOGY

VARIABILITY OF MORPHOLOGICAL CHARACTERS

If widely distributed and variously sympatric forms of crossbill are not made internally cohesive by gene flow, several factors might inflate levels of within-form morphological variation. If morphological characters are polygenic, quantitative, and heritable, as suggested by studies of other wild songbirds (Boag and Grant 1978, Smith and Zach 1979, van Noordwijk et al. 1980, P. R. Grant 1983, Alatalo and Lundberg 1986, B. R. Grant and P. R. Grant 1989), interbreeding between forms might increase levels of within-form variance (Mayr 1942, Mayr et al. 1953, Schueler and Rising 1976, P. R. Grant 1986). Alternatively, observations of high morphological variance in pooled samples of vocally similar crossbills from widely separated geographic regions might imply within-form geographic variation. Additionally, because crossbills experience a wide variety of environmental conditions, including variation in conifer cone morphologies, breeding seasons, altitude, and climate, they may be especially subject to differential environmental modification of morphology (sensu James 1983).

Variation in external measurements. Vocal groups of red crossbills showed remarkable uniformity in levels of external character variation (Table 5 and Appendix D). Comparisons among within-group CVs showed almost no significant differences (*F*-ratios) within characters (see Groth 1990). Sexes were not consistently different in levels of within-group morphological variation. Wing length and cube root of body mass had consistently the lowest CVs within groups, consistent with other avian studies. Wing and tail feathers are known to experience considerable shortening due to wear (e.g., J. Davis 1961, R. E. Johnson 1977, Alatalo et al. 1984, Francis and Wood 1989), which could thereby inflate CVs in this study because specimens from different seasons were pooled. Nevertheless, wing length CVs were low and similar to other published values for local populations of songbirds.

As in other birds, some bill measurement variation can be explained by growth and wear (J. Davis 1954, 1961; Packard 1967; Gosler 1987; R. E. Johnson 1977). Lack of

Table 5. Coefficients of Variation (CVs) and Sexual Dimorphism in Forms of Red Crossbill

Character	CV[a] Males	CV[a] Females	Sexual Dimorphism[b]
Cube Root of Body Mass	2.30 (1.95–2.82)	2.48 (2.02–3.49)	2.41 (0.94–2.71)
Tarsus Length	2.98 (2.68–3.59)	2.75 (2.17–3.47)	1.29 (0.00–2.21)
Toe Length	3.52 (2.74–4.06)	3.59 (2.65–4.81)	2.05 (-0.31–4.23)
Wing Length	1.75 (0.86–2.61)	1.93 (0.95–2.62)	3.47 (1.19–4.75)
Tail Length	3.27 (1.96–4.97)	3.18 (2.54–4.28)	3.55 (0.53–5.77)
Bill Character Mean (n = 6)[c]	3.61 (1.61–6.07)	3.71 (0.81–9.04)	2.33 (-0.05–5.63)
Skeletal Character Mean (n = 23)[c]	3.27 (2.27–6.66)	3.36 (1.45–9.42)	1.85 (-1.65–12.92)

a. Values indicate means of CVs across forms (n = 7 forms for males, n = 6 for females). Numbers in brackets indicate the range of intragroup CV values.
b. Values indicate mean percentage of sexual dimorphism across forms, calculated as ([male mean minus female mean]/female mean) for each character. Negative values indicate that females had greater means than males.
c. Values indicate grand means (means of character means across groups); numbers in brackets indicate the range of all within-group values used to calculate the grand means.

normal tomial occlusion in *Loxia* allows unrestrained growth of bill tips unless they are worn down by foraging or bill-wiping. Wild populations of crossbills use conifers that vary in cone morphology and ripeness, seed size, and thickness of seed husks, and these differences may influence measurements on the keratinized parts of the bill. Green or unopened cones require more biting with bill tips than mature cones that have separated scales. An example of a specimen with a long and overgrown bill (Fig. 23A, aM332) was taken on the east side of the Cascade Mountains, Oregon, in June 1985, where crossbills were foraging almost exclusively on very mature and open cones of ponderosa pine (*Pinus ponderosa*). No long-billed specimens taken for this study were as aberrant as the extremely long-billed white-winged crossbill (*L. leucoptera*) described by West (1974). A short-billed specimen (Fig. 23B, aM337) was taken in the hills of east-central Arizona on 1 August 1985, where cones of ponderosa pines were green; many relatively short-billed specimens at this and other localities had their bills covered with pine sap from green cones. Other changes in bill wear included abrasion of the tomia (Fig. 23C, aM542). The series of crossbills (which included aM542) taken in the Chiricahua Mountains, Arizona, in May of 1987 were feeding on thick-coated seeds of Apache pines (*P. englemannii*), and most birds had noticeable wear marks or grooves on both mandibles, apparently from seed husking. This source of variation might affect the measurement of upper mandible depth and also the width and depth of palatal grooves.

Variation in skeletal characters. Within-group CVs for skeletal characters were in the same range of values for external measurements (Table 5 and Appendix D). Furcular process length was the single most variable character within groups with CV values from 6–9%; furcular processes of crossbills are thin plates of bone that become thin at the edges and vary extensively among individuals in size and shape. Widths of long bones (coracoid, humerus, femur, and tibiotarsus) were also highly variable in all groups. Bone lengths, skull depth, and postorbital width were among the lowest in within-group CVs. Like external measurements, skeletal characters showed no consistent sexual or between-group differences in level of variability.

Variance due to leg asymmetry. Knox (1983) found that tarsi averaged shorter on the side of the bird to which the lower mandible crossed in *L. curvirostra, L. pytyopsittacus*, and also the Hawaiian Akepa *(Loxops coccinea)* which has a slightly-crossed bill. He suggested that increased strain on the leg opposite the direction of lower mandible crossing influenced "handedness." Therefore, within-population variation in leg characters may become inflated when leg bones (right or left) measured are not made consistently with respect to bill crossing direction. Handedness may be better described using a series of skeletal characters than with tarsus length alone (used by Knox 1983). Tarsus length on study skins is notoriously subject to measurement error in studies of small birds, whereas skeletal characters can be measured with more precision because proper placement of caliper tips is less ambiguous.

Only four of the nine leg bone characters showed trends in the direction consistent with Knox's (1983) findings, but none was significant (Table 6). Five characters (femur and tibiotarsus distal end widths and the three tarsometatarsal measurements) showed trends in direction opposite Knox's prediction, which was surprising because tarsus length is

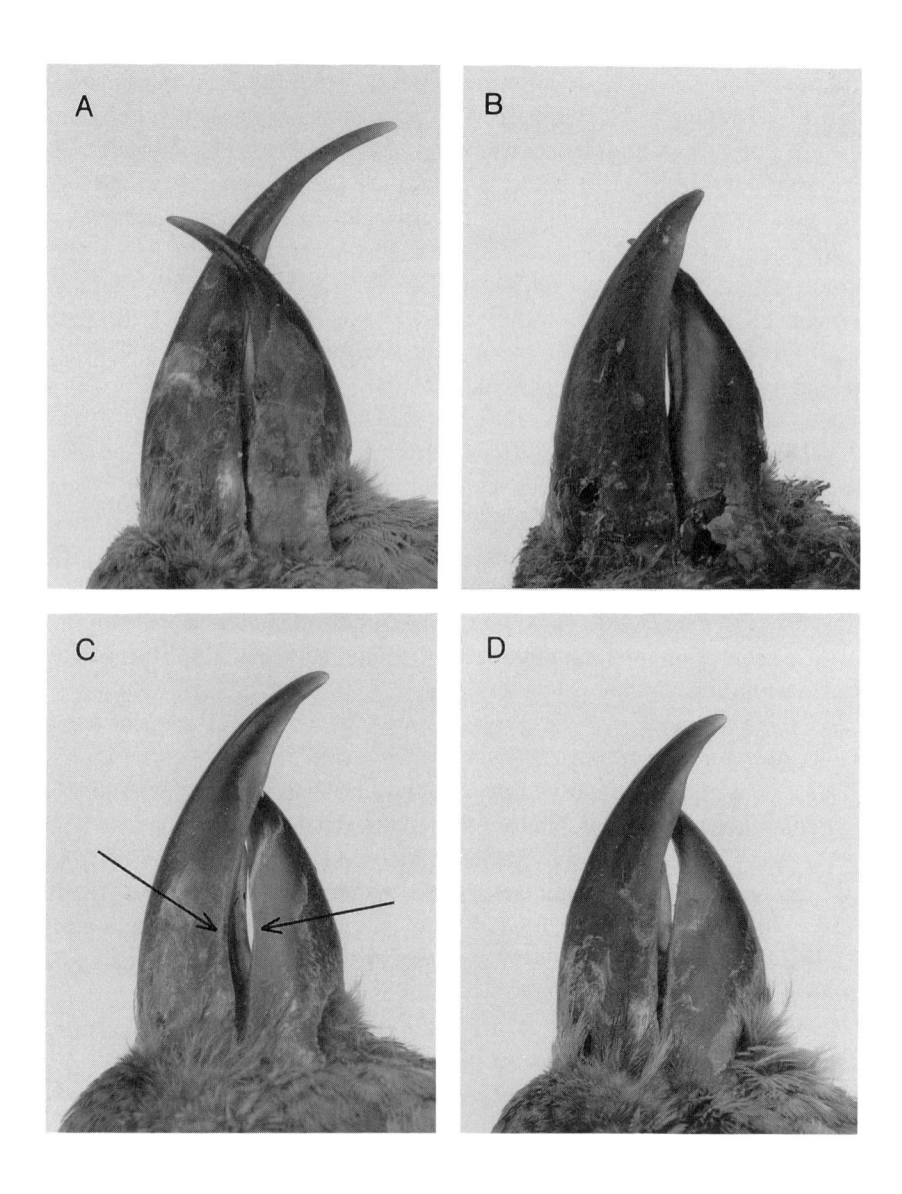

Figure 23. Variation in bill shape with wear. (A) Long and overgrown bill of aM332. (B) Short, worn bill of aM542. (C) Tomial wear on bill of aM542; arrows show areas with extensive wear. (D). Normal bill morphology of aM405.

Table 6. Leg Asymmetry in Skeletons of *Loxia*

Measurement	N	Mean (SE)[a]	min[a]	max[a]	t[b]	P[b]
Femur Length	35	-0.013 (0.015)	-0.28	0.12	-0.87	0.3928
Femur Width	35	-0.008 (0.004)	-0.05	0.05	-2.12	0.0415
Femur Distal End Width	37	0.011 (0.012)	-0.16	0.14	0.94	0.3530
Tibiotarsus Length	29	-0.067 (0.027)	-0.55	0.15	-2.50	0.0184
Tibiotarsus Width	32	-0.004 (0.004)	-0.05	0.14	-1.03	0.3122
Tibiotarsus Distal End Width	38	0.002 (0.007)	-0.07	0.13	0.25	0.8043
Tarsometatarsus Length	19	0.007 (0.017)	-0.12	0.18	0.43	0.6743
Tarsometatarsus Width	24	0.017 (0.006)	-0.07	0.05	0.30	0.7648
Tarsometatarsus Distal End Width	23	0.028 (0.010)	-0.07	0.12	2.93	0.0077

a. Difference between leg on the side to which the lower mandible crossed and the opposite leg (in mm). Negative values indicate that smaller measurements were on the side to which the lower mandible crossed.
b. Two-tailed paired *t*-test for significance of departure of mean difference from zero.

basically a measure of the length of the tarsometatarsus. Distal end width of the tarsometatarsus showed a highly significant trend opposite Knox's (1983) findings, but chance alone could cause a single significant difference among several characters. Furthermore, CVs in femur and tibiotarsus lengths did not exceed comparable non-leg skeletal characters of the same magnitude, such as coracoid length or humerus length (Appendix D; see Groth 1990). It seems warranted to conclude that leg asymmetry has no impact on character variance, at least for North American *Loxia*.

MORPHOLOGICAL GEOGRAPHIC VARIATION IN A SINGLE VOCAL GROUP

Character correlations (Table 7) within the Type 2 sample, which was taken from widely separated regions of North America, were reduced relative to correlations in heterogeneous samples of museum specimens (see Table 2). Low correlations indicate that less of the within-group variance can be considered "size" than in mixed samples. PC1 axes within Type 2 (Table 8) explained less than half of the total standardized variance but showed positive coefficients for all characters, indicating that they reflect general size to some extent. PC2 axes for the two sexes were similar in that they had high and positive loadings for upper and lower mandible lengths and low and/or negative loadings for body mass as well as widths and depths of mandibles. This allows the interpretation that PC2s measured bill length relative to bill thickness and body size.

For both males and females, regional samples overlapped greatly in morphology and revealed no geographic trends (Fig. 24). Type 2 crossbills fell within the same narrow range of morphology regardless of geographic origin. Local or regional variation in morphology was also not evident in the smaller samples of other forms which were also collected widely in different regions, which agrees with the low within-group variation seen in all forms.

SEXUAL DIMORPHISM

Measurement of sexual dimorphism in populations of *L. curvirostra* may be confounded because vocally and morphologically distinctive forms of crossbills often occur at the same sites in varying ratios. Unless sex ratios of the sampled forms are precisely equal, the within-sex means could be influenced by differences between forms. In this study, within-sex samples divided on the basis of vocalizations had low levels of morphological variation, suggesting that mixing of forms did not influence the following estimates of sexual size differences.

External measurements. Within vocal groups, males and females overlapped broadly in all external measurements (see Appendix D). The Type 7 samples were not large enough to allow estimation of mean sexual differences, but the single Type 7 female (aF497) was similar to the five Type 7 males. In the other six forms, males averaged slightly larger than females (Table 5). Within forms, males were 1.67–3.47% larger than females in mean dimorphism among the six bill size characters. Wing length showed highly significant sexual differences in all forms (except Type 6, which had small sample sizes), even though

Table 7. Correlation Coefficients Among Field Measurements in a Single Form ("Type 2") of Red Crossbill

					Characters				
	1	2	3	4	5	6	7	8	9
1. Cube Root of Body Mass	—	0.39	0.40	0.13	0.20	0.54	0.44	0.36	0.48
2. Tarsus Length	0.18	—	0.18	-0.06	0.07	0.34	0.33	0.34	0.24
3. Wing Length	0.23	0.15	—	0.29	0.23	0.32	0.24	0.17	0.17
4. Upper Mandible Length	0.06	0.24	0.17	—	0.67	0.30	0.33	0.24	0.10
5. Lower Mandible Length	0.03	0.18	0.16	0.57	—	0.31	0.35	0.31	0.08
6. Bill Depth	0.32	0.23	0.09	0.15	0.23	—	0.88	0.57	0.47
7. Upper Mandible Depth	0.29	0.19	0.01	0.16	0.16	0.83	—	0.50	0.41
8. Upper Mandible Width	0.15	0.12	0.08	0.24	0.19	0.37	0.30	—	0.37
9. Lower Mandible Width	0.25	0.08	-0.02	-0.04	0.05	0.27	0.15	0.11	—

a. Values for females (n = 117) above the diagonal; males (n = 188) below.

Table 8. Correlations (Factor Loadings) Between Measurements and the First Three Principal Components in a Single Form ("Type 2") of Red Crossbill

Variable	Males (n = 188)			Females (n = 117)		
	PC1	PC2	PC3	PC1	PC2	PC3
Cube root of body mass	0.491	-0.296	0.550	0.698	-0.288	0.339
Tarsus length	0.450	0.192	0.292	0.475	-0.461	0.126
Wing length	0.275	0.305	0.687	0.479	0.136	0.776
Upper Mandible length	0.492	0.695	-0.122	0.478	0.769	0.010
Lower Mandible length	0.505	0.629	-0.163	0.522	0.690	-0.046
Bill depth	0.829	-0.346	-0.198	0.878	-0.101	-0.187
Upper Mandible depth	0.765	-0.353	-0.275	0.833	-0.028	-0.275
Upper Mandible width	0.550	0.013	-0.248	0.694	-0.075	-0.312
Lower Mandible width	0.330	-0.405	0.192	0.592	-0.353	-0.067
Total variance[a]	30.0%	16.7%	12.4%	41.6%	16.9%	10.6%

a. Percentages of total standardized variance explained by each axis.

the mean differences were less than or comparable to other characters. Unlike size characters, curvature angle of the bill showed no consistent sexual dimorphism.

Skeletal measurements. Levels of sexual size dimorphism in skeletal measurements were similar to those for external measurements (see Table 5). Mean differences among characters within forms ranged from 0.74–4.14%, with males consistently larger than females, and only a few within-group comparisons in single characters showed females averaging larger than males. The seven characters of the pelvic and leg regions (synsacrum, femur, and tibiotarsus measurements) showed the lowest levels of dimorphism, whereas characters of the pectoral region (especially the furculum, sternum, and scapula) showed the highest dimorphism (see Groth 1990). Furcular process length, the character with the highest within-sex coefficients of variation (see preceding section), showed the highest overall levels of sexual dimorphism and ranged up to a nearly 13% mean difference in Type 4 birds.

VARIATION IN SIZE WITH AGE

The contribution of age differences to morphological size variation is dubious in passerine songbird populations. Apparently, the rapid and highly determinate development of adult size generally found in passerine birds precludes age differences (Ricklefs 1968, P. R. Grant 1981). Nevertheless, ornithologists sometimes exclude "first year" individuals from morphological analysis because of the possibility that young birds are smaller. Selander and Johnston (1967) found that first-year house sparrows (*Passer domesticus*)

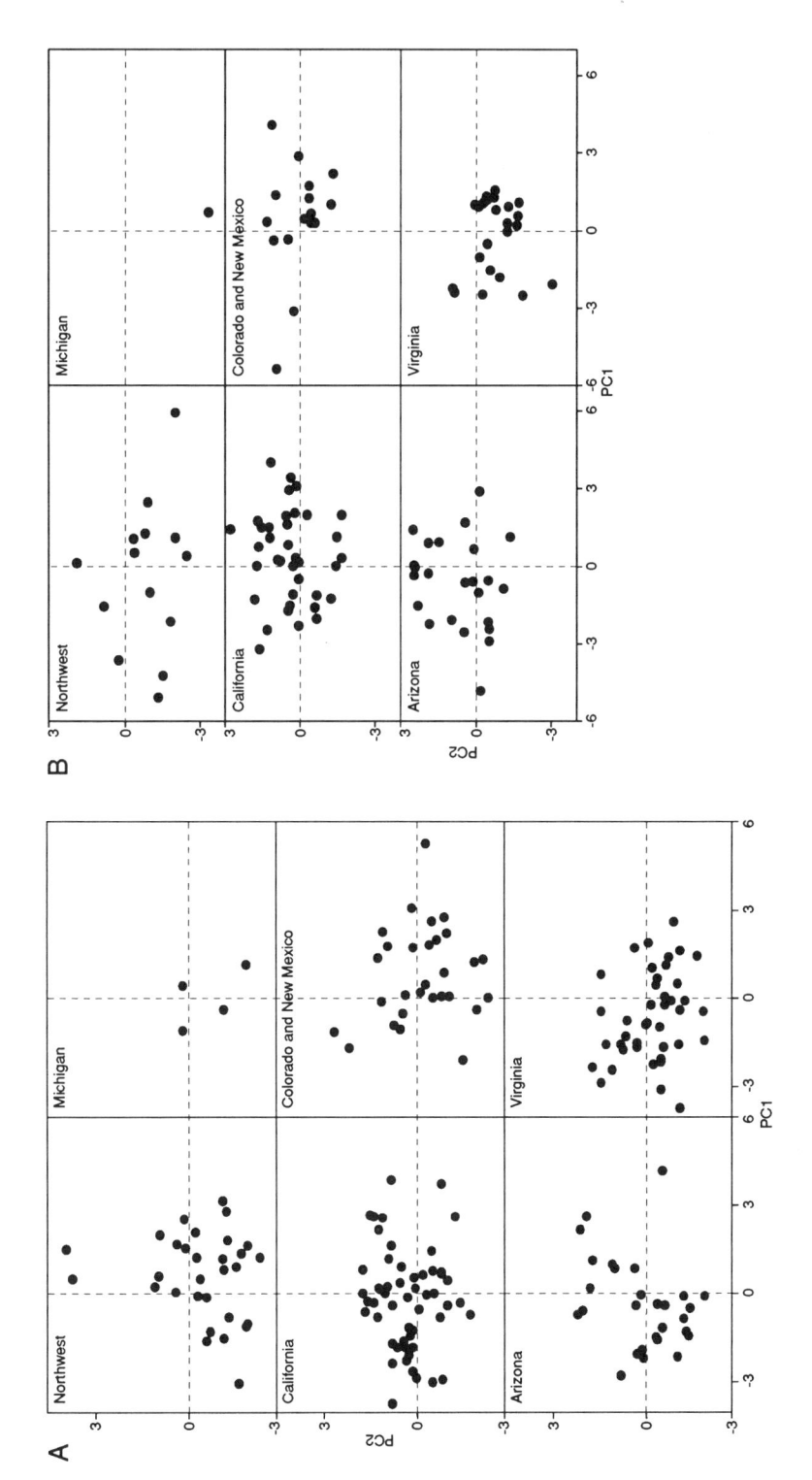

Figure 24. Principal components plots based on nine morphological measurements for six geographic subsamples of Type 2 males (A) and females (B).

averaged 0.08–1.41% smaller than adults over a set of linear external measurements. Differences of this magnitude are important because differences between populations of crossbills can be less than 5% (see next chapter).

This study is the first to divide crossbills into morphologically homogeneous and vocally distinctive populations. Tordoff (1952) divided a sample of *L. curvirostra* into "immature" and "adult" subsamples and compared them in four external size characters. All birds had solid-colored body plumage and would have been considered "adult" in the present study; Tordoff used cranial pneumatization and colors of feather edges as age criteria. He found that adults were slightly larger than immatures in all four characters, but only the difference in wing length was significant. He admitted that he had pooled the individuals of two subspecies differing in size, *L. c. bendirei* and *L. c. benti*, in his analysis. This suggests that any biases in proportions of the two subspecies could have influenced the results; specifically, a higher frequency of the smaller *bendirei* may have been included in the "immature" sample.

Correspondence between skull pneumatization and plumage maturation. The usefulness of plumage characters and cranial pneumatization in aging crossbills has been debated. McCabe and McCabe (1933:136) found a streaked male showing "more doubling [of the cranium] than several of the pure red males and greenish-yellow females," which is among the first suggestions of discordance between plumage and skeletal development in crossbills. The McCabes (p. 136) also wrote that in their *L. curvirostra* "the foreparts of the frontal bones of the skull never assume the usual two-layered adult passerine character." Phillips (1977:113) disagreed with the last statement, but did confirm that birds with adult plumages could have skulls not fully pneumatized, saying "such cases may result, in part, from a protracted rate of skull ossification." Tordoff's (1952:200) specimens were all beyond the postjuvenal molt, yet he also found that "in many of the specimens which I collected in November and December, 1950, the skull was incompletely ossified. By the middle of January, 1951, cranial ossification seemed to be complete in all specimens collected." This observation suggests that the timing of pneumatization in crossbills is similar to that in other migratory North American songbirds, in which complete pneumatization requires about six months (A. H. Miller 1946, Chapin 1949, Wood 1969, Yunick 1977). Other songbirds show slower rates of pneumatization or may never reach complete double-layering (Selander 1958, 1964; Bowman 1961).

Plumage development and cranial pneumatization are only weakly synchronous in crossbills (Fig. 25). No birds with plumage scores less than 6 had fully pneumatized crania, but many had over 60% completion. Many streaked juveniles with only a few flecks of color had only very small single-layered areas. On the other hand, 16 specimens with no juvenal contour feathers (score 9) had incomplete pneumatization. All of these could have been less than one year old based on color of the edges of remiges and wing coverts (Ticehurst 1915, Phillips 1977), but because these characters are inconsistent, some may have been older. Nevertheless, the low frequency of such birds suggests that it is unlikely that crossbills retain single-layered areas of the skull throughout life, as was suggested by McCabe and McCabe (1933).

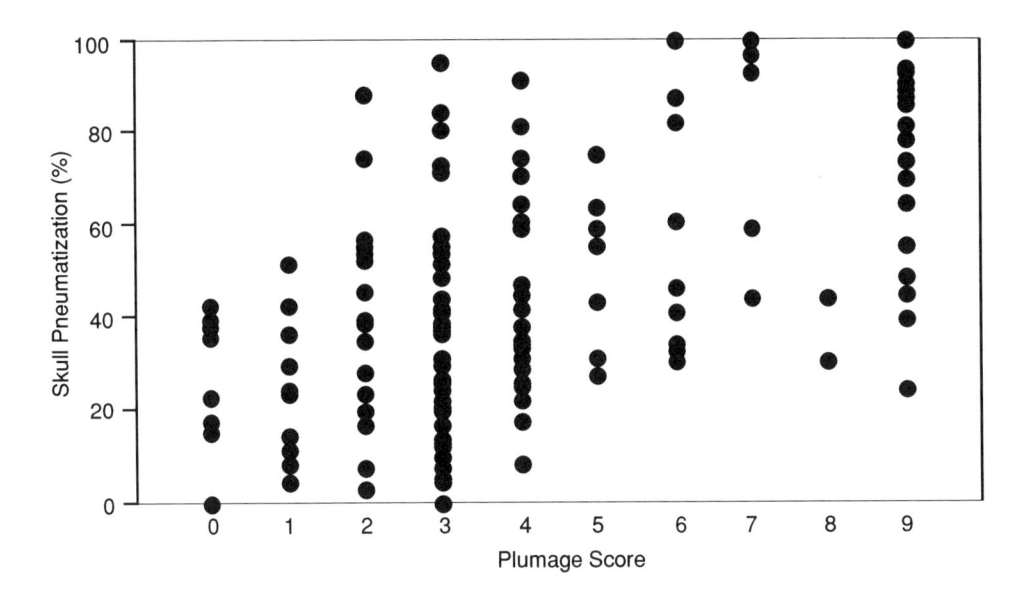

Figure 25. Correlation between plumage score and skull pneumatization. The data point at plumage score 9 and skull pneumatization 100% represents several hundred individuals.

It is evident that plumage development and skull pneumatization correspond only roughly. The explanation favored here is that molt is more seasonal (Dwight 1900, Phillips 1977), whereas pneumatization proceeds at a more constant rate. For example, juveniles hatched in the spring would accumulate only a few adult contour feathers (plumage scores 0 to 3) through the summer before a more rapid and complete molt in autumn, with skull pneumatization nearing completion in many birds before the molt. On the other hand, birds hatched in late summer would rapidly molt from juvenal to adult plumage a few weeks after leaving the nest, with the skull becoming double-layered some time later. This suggests that age estimation is more accurate using pneumatization; unfortunately, skulls of most specimens in museums are not visible. No evidence was found for a protracted rate of pneumatization as Phillips (1977) suggested.

Size and plumage development. The extent of growth of juvenile specimens was estimated by dividing their measurements by the adult (plumage score 9) character means in groups based on vocalizations and sex to obtain percentage values. These values were then pooled for all groups and divided into "age" categories by plumage score. The spread of points at plumage score 9 (Fig. 26A) represents the within-population distributions among adults. Wing length and humerus length in birds with plumage scores 1 to 8 were not different from adults, suggesting that these characters reach full size only a few weeks after fledging. Bill depth also showed a rapid increase, but bill length did not reach the adult mean in plumage classes less than 6. Other skeletal characters (not shown) showed the same pattern of rapid development as humerus length.

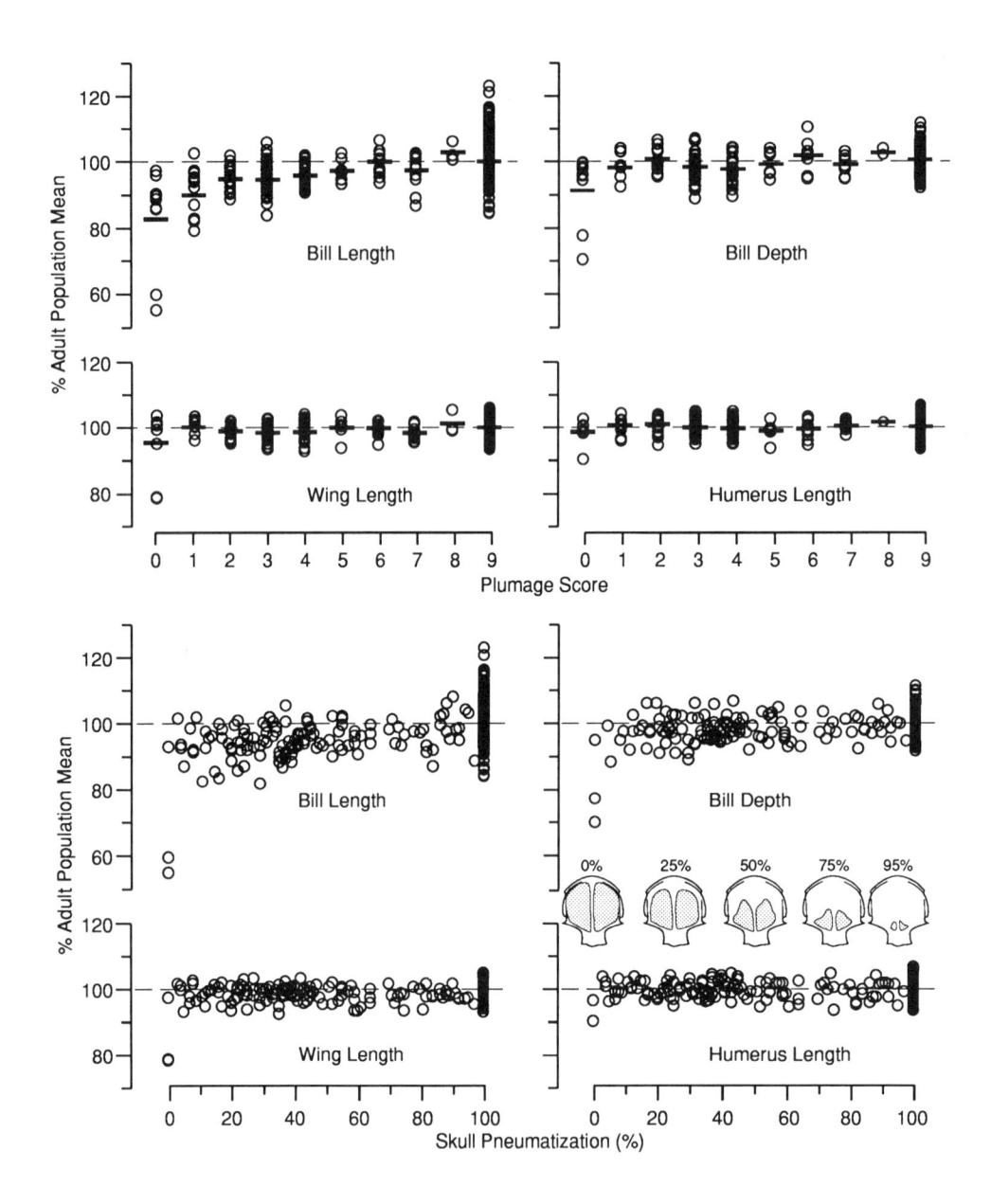

Figure 26. Size development with plumage score (top; horizontal bars represent within-class means) and pneumatization values (bottom) for four characters in immature red crossbills. Distributions at plumage score 9 and skull pneumatization 100% represent adult populations.

Skull pneumatization and size. The percentage of adult size for juveniles was estimated as described above, except that "adults" were considered those with 100% pneumatized skulls (therefore adult means had miniscule differences from those in the plumage analysis above). Humerus length and wing length reached the plateau of full adult size long before pneumatization was complete (Fig. 26B); only the very youngest juveniles (0% pneumatized) were small. Bill depth developed only slightly less rapidly than wing and humerus lengths. However, most birds with incomplete pneumatization had shorter bills than adults of their respective populations.

DISCUSSION

Morphometric character variation within vocal groups of red crossbills was low, even though most samples were aggregates of specimens collected at geographically dispersed sites. For example, within-form CVs of wing length averaged 1.8%, which is lower than the average within-population CV of 2.3% (range 1.4% to 3.2%) in the 37 passerine taxa I listed in a previous study (Groth 1988). Furthermore, even though characters such as upper mandible length are subject to extensive abrasion and wear, the mean CV of 4.6% in crossbill vocal groups is in the middle of the distribution of bill length CVs in the same set of passerine taxa. CVs of skeletal characters were also low, but ranged more widely than external characters (up to 9.4% in furcular process length), but long bone lengths generally had CVs averaging 2–3%. These results do not support the hypotheses of hybridization between crossbill forms, within-form geographic variation, or differential environmental modification of morphology. Low morphological variance supports the hypothesis that red crossbill forms represent separate populations made internally cohesive by gene flow.

Sexual dimorphism was an important source of intrapopulation variation. Crossbill forms had similar levels of sexual dimorphism, with males consistently about 2% larger than females. The finding of a ubiquitous level of dimorphism suggests that sexual dimorphism was present in the ancestor and has remained unmodified through the evolution of different crossbill forms. As suggested by Price (1984) for similar findings in Darwin's finches, selective forces responsible for the observed dimorphism, if any, may have operated in the distant past. It is not known if sexual selection continues to operate to maintain the dimorphism. Although male and female crossbills forage in the same trees, bill characters did not show lower levels of sexual dimorphism than other characters, as was found in house sparrows (*Passer domesticus*) by Selander and Johnston (1967).

Age variation contributed little to variation in size. Therefore, not only first-year but also many young, streaked crossbills can be included in morphometric samples without influencing estimates of within-population size variation. Generally, museum specimens with at least a single adult contour feather (plumage score 1 and above, Table 1) could be pooled in morphological analysis. The only recommended corrections for these juveniles are for upper and lower mandible lengths, for which an addition of 1% of the raw measurement per plumage score unit less than 9 would give approximate predicted adult values.

MORPHOLOGICAL DIFFERENCES AMONG VOCAL GROUPS

The North American *L. curvirostra* complex exhibits size variation as a continuous gradient, therefore previous divisions of the complex involved arbitrary morphological limits. This chapter examines the significance of divisions of red crossbills based on vocalizations. The degree to which morphology is different among groups will indicate the taxonomic utility of vocal differences. Here, patterns of both univariate and multivariate differentiation among vocal groups are described.

UNIVARIATE PATTERNS

External and skeletal measurements. Adult crossbills, divided into samples based on call note structure, showed consistent patterns of univariate differentiation (Appendix D). Type 6 birds were largest for all characters, followed by Type 2, Type 5, and Type 7, which were second-largest, depending on the character. Type 1 and Type 4 were similar to each other in character means and distributions, and both were smaller than all other groups except Type 3. Type 3 had the smallest character means which were, for both sexes, significantly different (ANOVA, $P < 0.05$) from the next-to-smallest means for each character, which were from either Type 1 or Type 4. The magnitude of size difference between some forms was slight (Fig. 27).

Appendix D contains only the field measurements for six characters that were also measured on dried study skins. The average percentages of shrinkage for these characters were: wing length, 1.1%; upper mandible length, 0.2%; lower mandible length, 0.2%; bill depth, 1.2%; upper mandible width, 4.3%; lower mandible width, 2.9%.

Wing size and shape. The major differences among wings of the vocal groups were in absolute size, not wing shape (Fig. 28). Male and female samples were arranged in the same relative sequence of wing size. Type 6 wings consistently averaged largest, followed by Type 5. Type 5 wing feathers averaged longer than those of Type 2, and sample sizes were large enough so that the mean differences between Type 2 and Type 5 were highly significant (two-tailed *t*-tests, all P values < 0.0016) for all feather lengths, although

Figure 27. Study skins of seven vocally defined groups of red crossbills. Representatives are adult males with measurements of bill length and bill depth near their population means, and are arranged from top to bottom in order of increasing mean bill size in populations, as follows: Type 3 (aM272); Type 1 (aM143); Type 4 (aM304); Type 7 (aM706); Type 5 (aM406); Type 2 (aM243); Type 6 (aM747).

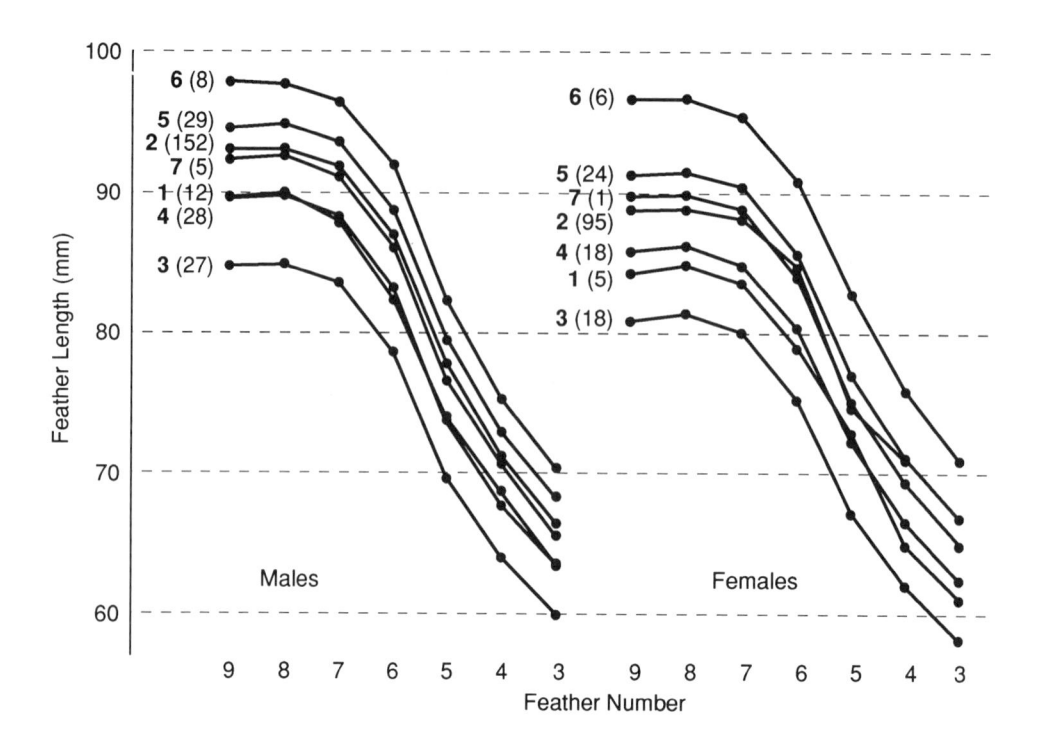

Figure 28. Wing size and shape in seven groups of red crossbills. Data points show the mean values for the seven outer primary feathers for each group. Call type numbers are indicated in boldface; sample sizes are in parentheses.

overlap was extensive between these two populations. Wings of Type 7 birds were close to those of both Type 2 and Type 5, but small samples prohibited satisfactory estimates of Type 7 means. Type 1 and Type 4 showed no significant differences from each other in feather lengths in samples of both males and females (two-tailed t-tests), although Type 4 mean feather lengths were larger than those of Type 1 for most comparisons. Both Type 1 and Type 4 had mean lengths significantly smaller than Type 2 and significantly larger than Type 3. Males showed no obvious differences from females in wing shape; sexual differences within forms were mainly in size.

Bill curvature. Males and females within vocal groups did not show significant differences in curvature angle, therefore the sexes were pooled for comparisons between groups. Distributions in curvature angle overlapped greatly among the seven forms (Fig. 29) and showed less differentiation than linear size measurements. Type 6 birds stood out as having the most highly curved bills, and the Type 6 population mean was significantly different (two-tailed t-tests, $P < 0.05$) from the other samples. Differences among the other groups were not significant.

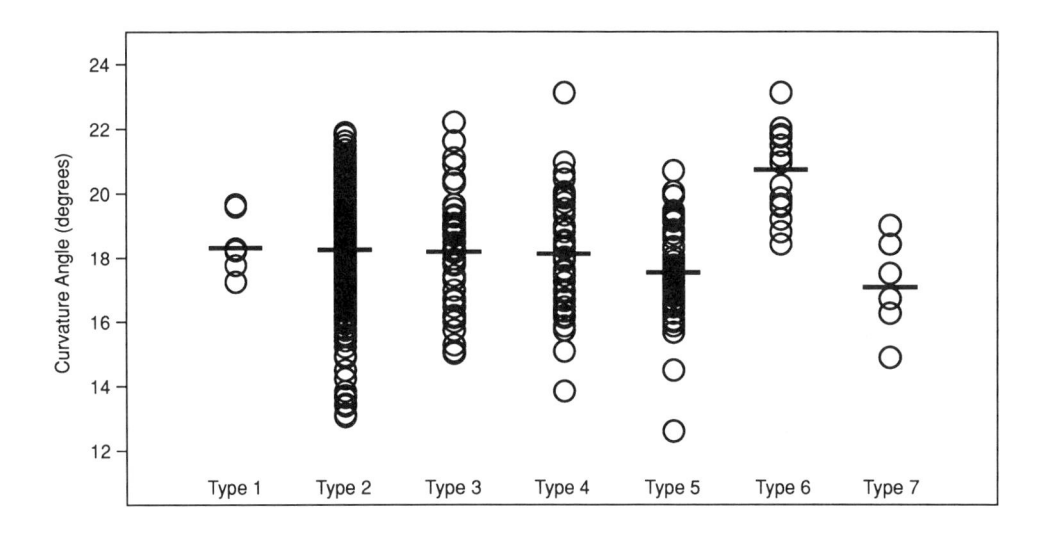

Figure 29. Bill curvature in seven groups of red crossbills. Circles represent single individuals; horizontal bars indicate the mean for each group. See Appendix D for sample sizes.

PRINCIPAL COMPONENTS ANALYSIS

Bill characters. In the samples with all adults pooled, correlation coefficients among bill characters (n = 15 for each sex) were high and positive. Correlation coefficients for males averaged 0.776 and ranged from 0.67 (lower mandible width/lower mandible length) to 0.96 (bill depth/upper mandible depth); coefficients for females averaged 0.769 and ranged from 0.64 (upper mandible length/lower mandible width) to 0.95 (bill depth/upper mandible depth). See Groth (1990) for complete correlation matrices.

About 81% of the sum of the standardized variances of the six bill measurements could be expressed as single factors in both sexes (Table 9). All bill characters were correlated to PC1 axes with coefficients of at least 0.85; therefore, PC1 axes in both sexes can be interpreted as reflecting bill size. PC2 vectors had similar loadings in the two sexes, with the two "bill length" characters loaded positively and the four "bill thickness" characters loaded negatively, so that birds with high and positive PC2 scores had relatively long, narrow bills. Although PC3 axes accounted for only minor fractions of the variances in both sexes, the sexes were highly similar in character loadings on PC3, revealing similarity in morphological pattern between the two sexes.

Plots of PC1 versus PC2 (Fig. 30) show that vocally defined forms of crossbill occupy narrow limits within the total range of bill size variation. Differences among forms were mainly along PC1 (bill size), even though computation of PC1 was not constrained to follow the trajectory of maximal variance between group means. Little differentiation among groups was apparent along PC2 (relative bill thickness).

Table 9. Correlations (Factor Loadings) Between Bill Measurements and the First Three Principal Components

Variable	Males (n = 329)			Females (n = 218)		
	PC1	PC2	PC3	PC1	PC2	PC3
Upper Mandible Length	0.872	0.404	0.006	0.866	0.421	0.086
Lower Mandible Length	0.862	0.432	0.020	0.878	0.385	0.041
Bill Depth	0.951	-0.212	-0.115	0.947	-0.196	-0.140
Upper Mandible Depth	0.936	-0.211	-0.193	0.935	-0.175	-0.207
Upper Mandible Width	0.918	-0.146	-0.116	0.907	-0.093	-0.150
Lower Mandible Width	0.878	-0.218	0.427	0.855	-0.315	0.411
Total Variance	81.6%	8.5%	4.1%	80.8%	8.4%	4.4%

a. Percentages of total standardized variance explained by each axis.

Skeletal characters. With all individuals within each sex pooled, all correlation coefficients among skeletal characters (n = 210 per sex) were positive but generally not as high as among bill characters (see Groth 1990). Correlation coefficients for males averaged 0.515, ranging from 0.12 (coracoid width/furcular process length) to 0.90 (humerus length/femur length); coefficients for females averaged 0.549 and ranged from 0.11 (coracoid width/synsacrum minimum width) to 0.91 (humerus length/femur length).

Character loadings on the first three principal component axes were similar in direction and magnitude for both sexes (Table 10). PC1 axes can be described as body size vectors because all loadings were high and positive, and in both sexes explained about 58% of the sum of the standardized variances among characters. On PC1 axes, bone widths generally showed slightly lower loadings than bone lengths, probably reflecting increased measurement error and/or larger within-population variances in these characters. PC2 axes for both sexes had widths of long bones (coracoid, humerus, femur, and tibiotarsus) loading most highly and positively, whereas bone lengths and skull characters loaded negatively, allowing the interpretation that PC2 axes were reflections of relative long bone thickness. Although bone widths generally have higher levels of measurement error than bone lengths in studies of avian skeletons, the observation that all long bone widths loaded in the same direction on PC2 axes indicates that bone widths form a correlated subset of characters. For both sexes, PC3 axes had furcular process length loading most highly, with other character loadings fluctuating around zero.

Table 10. Correlations (Factor Loadings) Between Skeletal Measurements and the First Three Principal Components

Variable	Males (n = 262)			Females (n = 170)		
	PC1	PC2	PC3	PC1	PC2	PC3
Postorbital Width	0.746	-0.100	-0.188	0.752	-0.203	0.058
Skull Width	0.875	-0.097	-0.093	0.720	-0.169	0.118
Skull Depth	0.822	-0.057	-0.144	0.817	-0.055	-0.065
Coracoid Width	0.382	0.698	0.022	0.456	0.759	-0.123
Coracoid Length	0.898	-0.072	-0.044	0.915	-0.074	-0.156
Scapula Width	0.773	-0.011	-0.042	0.786	0.048	-0.005
Scapula Length	0.872	-0.105	0.039	0.897	-0.060	-0.089
Humerus Width	0.740	0.356	0.045	0.781	0.052	0.126
Humerus Length	0.890	-0.109	-0.091	0.904	-0.054	-0.240
Sternum Length	0.883	-0.084	0.068	0.885	0.042	0.052
Keel Length	0.808	-0.090	0.109	0.830	0.093	0.032
Sternum Width	0.679	-0.229	0.070	0.651	-0.023	0.092
Keel Depth	0.741	-0.109	0.451	0.691	-0.011	0.104
Furcular Process Length	0.414	0.029	0.776	0.410	0.036	0.792
Anterior Synsacrum Length	0.781	-0.099	0.025	0.809	-0.054	-0.115
Synsacrum Minimum Width	0.615	0.008	-0.365	0.606	-0.347	0.215
Synsacrum Width	0.789	0.017	-0.066	0.804	-0.008	-0.021
Femur Width	0.689	0.435	-0.084	0.760	0.117	0.111
Femur Length	0.877	-0.119	-0.048	0.868	-0.121	-0.232
Tibiotarsus Width	0.623	0.533	0.002	0.631	0.507	0.122
Tibiotarsus Length	0.848	-0.190	-0.092	0.871	-0.096	-0.212
Total Variance	58.2%	6.1%	4.9%	58.8%	5.2%	4.7%

a. Percentages of total standardized variance explained by each axis.

Patterns of population relationships in skeletal morphology were similar for the two sexes (Fig. 30). The Type 6 population had the largest body size, but for both sexes the plots revealed some overlap with Type 2 individuals. Undoubtedly, larger samples of Type 6 birds would increase the amount of overlap with Type 2 and the two other large-bodied populations, Type 5 and Type 7. Type 2, Type 5, and Type 7 occupied nearly identical positions in skeletal morphospace. Type 1 and Type 4 were nearly identical in skeletal PC space, and both overlapped Type 2 and Type 5. Skeletons of Type 3 birds overlapped distributions of Type 1 and Type 4 populations, and overlap was particularly great in females. A few of the largest Type 3 females had skeletons large enough to overlap some of the smallest Type 2 females.

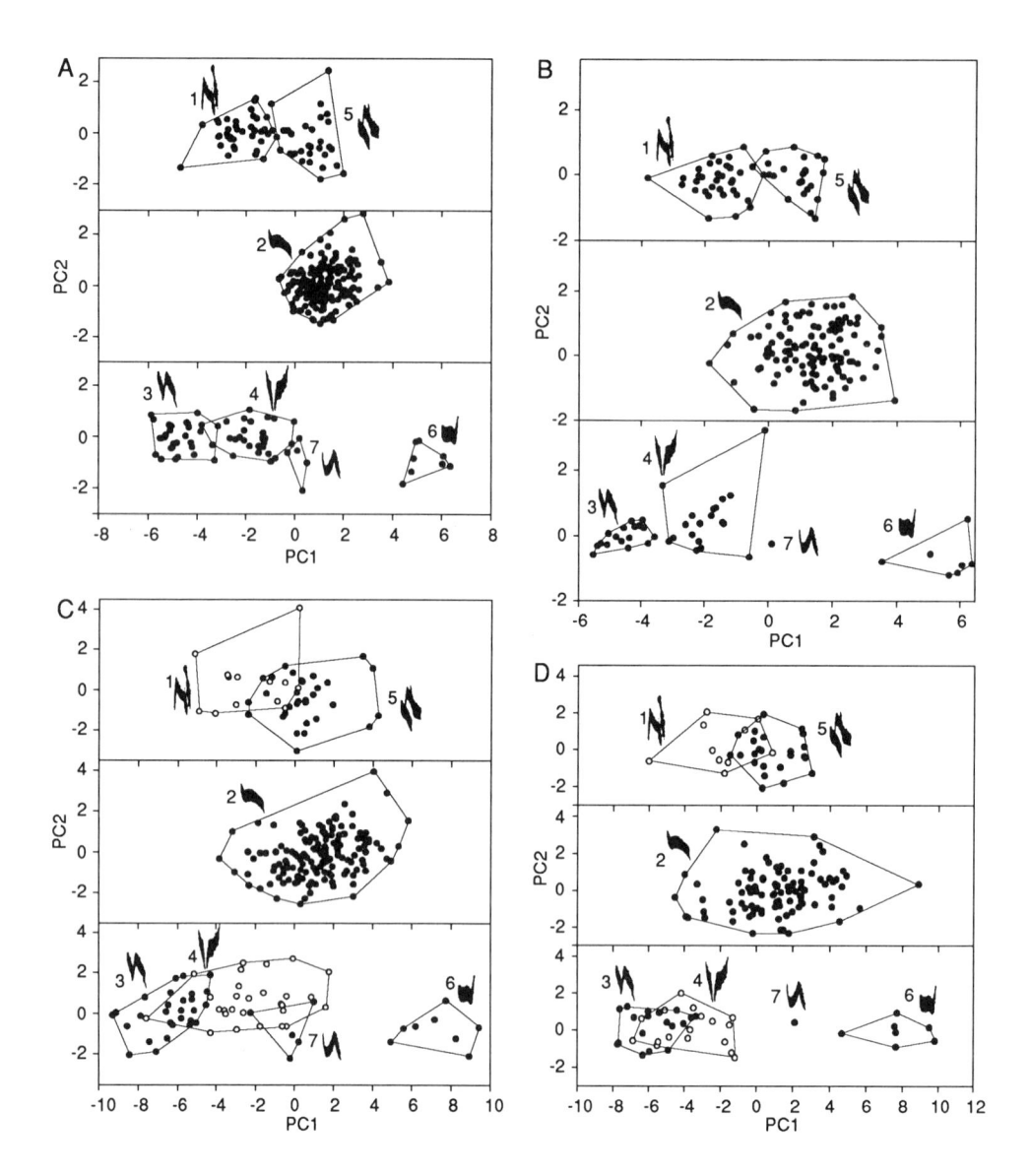

Figure 30. Principal components plots based on separate analyses of bill (A, males; B, females) and skeletal (C, males; D, females) characters. The three plots for each analysis portray the same morphospace. Minimum convex polygons are drawn around the peripheral members of each vocally defined group; numbers and representative audiospectrograms identify the 7 groups.

PCAs using skeletal characters did not separate forms as greatly as in PCAs using bill characters. For example, although Type 3 and Type 4 populations overlapped considerably in skeletal characters, they were nearly completely separated using bill characters. Type 1 and Type 4 showed nearly perfect coincidence for both bill and skeletal character sets, as did the subset consisting of Type 2, Type 5, and Type 7.

MORPHOLOGICAL DISCRIMINATION OF VOCAL GROUPS

When seven study skin measurements are examined using canonical discriminant analysis, groups of red crossbills showed a range of levels of multivariate morphological separation (Fig. 31). This method normalized the canonical coefficients so that the pooled within-group variance was 1; therefore, approximately 95% of all specimens in each group should fall within a radius of 2 standard deviations, or 2 units, of the corresponding group centroid. Empirically, the different groups had nearly identical coefficients of variation for univariate characters; therefore, it is not unreasonable to assume that the multivariate dispersions in each form were nearly the same. Evidence supporting this assumption was that the distributions (not shown) of the large samples of Type 2 males and females fell in circular clusters, and only a few data points were beyond 2 standard deviations of the centroids.

In analyses of both sexes, scores of group means along the first canonical axis (CAN1) were highly correlated with group means for the seven raw measurements, indicating that CAN1 axes were basically size axes (Table 11). Second canonical axes accounted for greatly reduced amounts of the total among-group variance, and had loadings (among-group correlations) indicating a contrast between bill measurements and wing and tail lengths. Specimens with long wings and tails relative to bill size received higher scores for CAN2 in both sexes. Canonical axes beyond the first two added little additional discriminatory power.

Type 6 specimens were more highly separable from specimens in the nearest neighboring groups (Types 2 and 5) than in other pairwise comparisons between adjacent groups. Type 3 males were more separable from adjacent groups (Types 1 and 4) than were female specimens. Crossbills of intermediate size (Types 1, 2, 4, 5, and 7) formed an extensively overlapping aggregation. Two-group discriminant analyses between pairs of vocal groups similar in overall size (Type 1 versus Type 4, and Type 2 versus Type 5) provided additional separation (see Groth 1990). In these analyses, the Type 1 and Type 4 groups showed only weak separation, but Type 2 and Type 5 specimens were separated largely on the basis of a contrast between wing and tail lengths and bill size.

In a canonical discriminant analysis using a combination of 32 external and skeletal characters, groups were only slightly more separable than in the analysis above using seven study skin characters (see Groth 1990). Mahalanobis' distances (Table 12) in this analysis provide a measure of relative separation among groups. With input of the same 32 characters into the SAS (1985) DISCRIM procedure, 92% of males and 94% of females were correctly classified into vocal groups (Groth 1990).

Table 11. Canonical Discriminant Coefficients for the Separation of 7 Forms of Red Crossbills Based on Study Skin Measurements

Variable	Males CAN1[a]	r[b]	CAN2[a]	r[b]	Females CAN1[a]	r[b]	CAN2[a]	r[b]
Upper Mandible Length	0.406	0.976	-0.308	-0.118	0.093	0.959	0.038	-0.174
Diagonal Upper Mandible Depth	0.670	0.993	-0.298	-0.038	0.555	0.990	0.412	-0.091
Upper Mandible Width	1.408	0.996	0.405	-0.050	0.942	0.993	-0.435	-0.051
Lower Mandible Width	1.032	0.991	-1.096	-0.106	0.908	0.987	-0.741	-0.064
Lower Mandible Length	0.066	0.984	-0.538	-0.159	0.450	0.952	-1.207	-0.273
Wing Length	0.304	0.974	0.286	0.222	0.207	0.977	0.283	0.205
Tail Length	-0.060	0.904	0.304	0.406	0.072	0.948	0.235	0.286
Constant[c]	55.71		21.74		47.45		16.36	
Total Variance[d]	93.1%		4.5%		88.5%		6.7%	

a. Canonical discriminant coefficients.
b. Correlation coefficients between population means on canonical axes and population means for raw measurements.
c. Constants to be subtracted from the sums of the raw measurements multiplied by their canonical discriminant coefficients.
d. Percentage of total among-group variance explained by each axis.

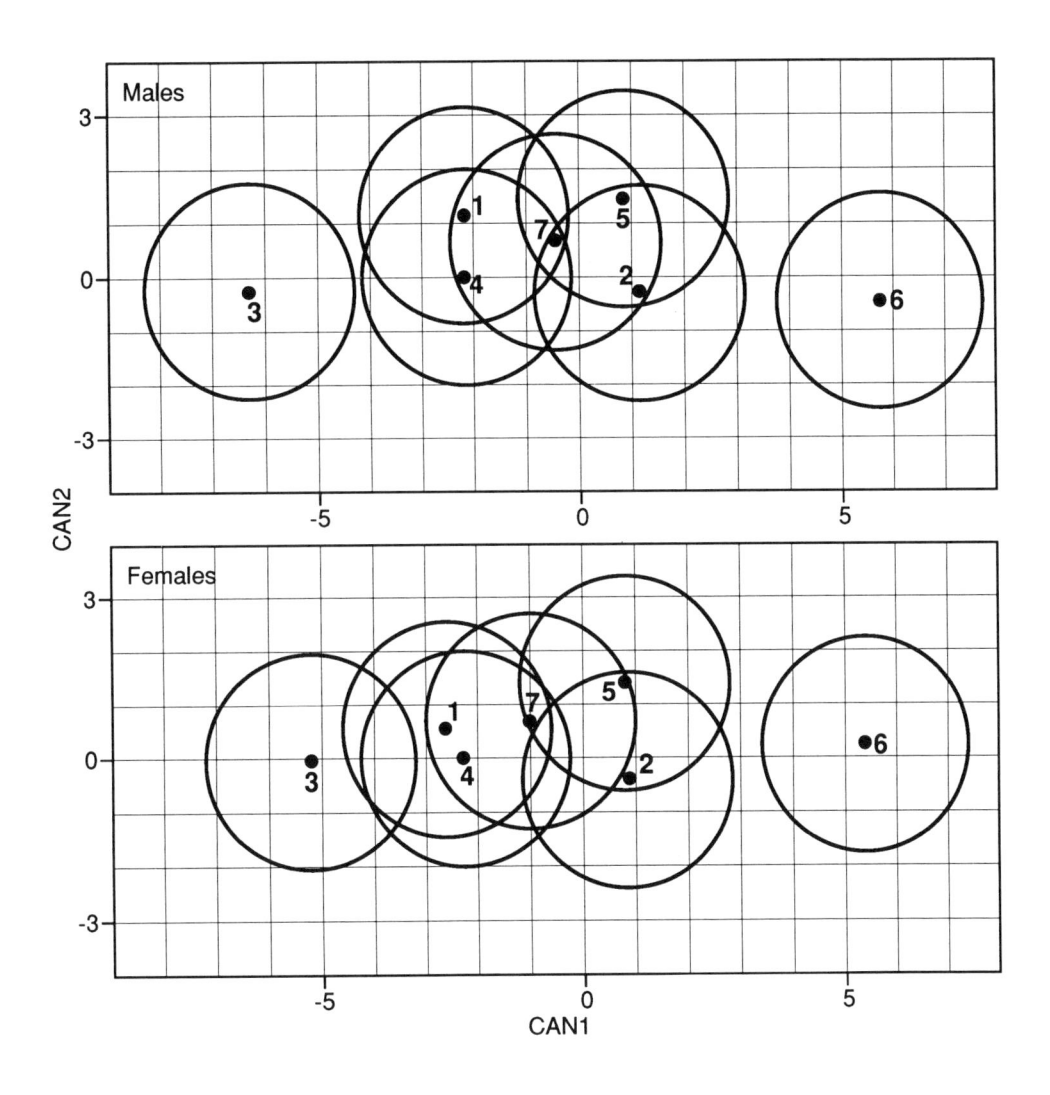

Figure 31. Canonical discriminant analysis plots based on 7 study skin measurements, with males and females in separate analyses. Numbered black dots indicate positions of group centroids for the seven vocally defined groups, and circles indicate areas within which approximately 95% of all specimens in each group are expected to fall.

Table 12. Mahalanobis' Distance Estimates, Based on Canonical Discriminant Analysis of 32 Morphometric Characters, Among 7 Forms of Red Crossbills[a]

	Type 1	Type 2	Type 3	Type 4	Type 5	Type 6	Type 7
Type 1	—	6.483	4.792	4.009	7.846	14.028	9.538
Type 2	6.207	—	7.809	4.334	2.748	8.713	7.304
Type 3	5.477	8.941	—	4.220	8.257	14.982	10.021
Type 4	3.394	4.484	5.070	—	5.141	12.280	8.455
Type 5	6.542	2.393	9.029	5.030	—	8.741	7.936
Type 6	12.606	7.085	15.243	11.170	7.511	—	10.805
Type 7	6.289	3.174	8.250	4.602	2.864	8.456	—

a. Distances for females above diagonal; males below.

SUMMARY

Vocalizations are predictive of morphology in North American red crossbills. Individuals of distinctive vocal populations occupied narrow zones within an otherwise continuous distribution of morphological variation. Morphologically intermediate individuals did not have call notes that were intermediate in structure. Vocally defined crossbill forms are most simply understood as discrete and equivalent units, each with equivalent and narrow ranges of morphology. Alternatively, morphology is only partially predictive of vocalizations. All of the smallest birds gave Type 3 calls, but slightly larger (but still small) birds gave either Type 1 or Type 4 calls. Although highly distinctive in calls, the Type 1 and Type 4 populations exhibited insignificant mean differences in morphology. Type 7 crossbills were slightly larger than those giving Type 1 and Type 4 calls, and there was extensive morphological overlap among Type 7, Type 2, and Type 5 birds. Even if entire suites of morphological characters are available for unknown specimens in this size range, vocalizations cannot be predicted with certainty for most specimens. Therefore, whereas calls predicted only one morphological class, morphology, in turn, predicted one or more vocal groups (as least within the present data set). Museum workers using study skin measurements will be able to separate most Type 3 and Type 6 specimens from other North American crossbills. Because forms of intermediate size overlap extensively, the vocal group to which specimens in this range correspond can be predicted with less certainty. Assignment of unknown specimens also depends on geographic information, which at this time is poorly known.

COLOR VARIATION

Griscom (1937) and Monson and Phillips (1981) suggested that different populations of crossbills show consistent color differences, and Massa (1987) showed "clinal" variation in the frequencies of yellow versus reddish males among European populations. These studies suggest that color may be useful in examining variation among crossbill forms.

Red crossbills are highly variable in plumage coloration. Males vary from yellow to orange to red, and shades of red vary from bright pinkish to scarlet to dull brownish red. Females range from gray to olive green to yellow-orange. The relationship of color to age has been debated. Wheelwright (1862:8001) suggested that in male crossbills and pine grosbeaks (*Pinicola enucleator*), "the red plumage is only an intermediate stage, and the full mature dress of both species is bright yellow green," which was consistent with the description by Linnaeus (1758) for the pine grosbeak, "*junior ruber, senior flavus.*" Wheelwright (1871) later suggested that molting into yellow by old males was an artifact of captivity, a phenomenon which has since been commonly found in species of cardueline finches that are normally red. However, the presence of yellow males in nature remains incompletely understood, and the present notion is that they are younger than fully red males.

Ticehurst (1915) attempted to separate crossbills into age classes based on plumage characters. He suggested that the amount of red or yellow was not related to age, and indicated that mixtures of red and yellow contour feathers were found on nonmolting individuals. He also suggested that males molt all of the juvenal plumage in the first autumn except wing, tail, primary coverts, and the outermost greater coverts, and that buffy edges on the coverts were present only on first-year birds. Griscom's (1937) descriptions of subspecific variation in North American crossbills were based largely on shades of red in males; yellow birds were not considered adult or were discarded as "xanthochroistic" variants. Tordoff (1952) later showed that many males molt directly from the streaked to the red plumage, and that many fully red birds had incompletely pneumatized skulls, showing again that color was not a good criterion for age. Jollie (1953) disagreed with Tordoff and proposed a system of four age/plumage stages for populations in Idaho, in which males molted from the streaked juvenal plumage to the "mottled orange-yellow of the first immature" (p. 193). In Jollie's system, young yellow

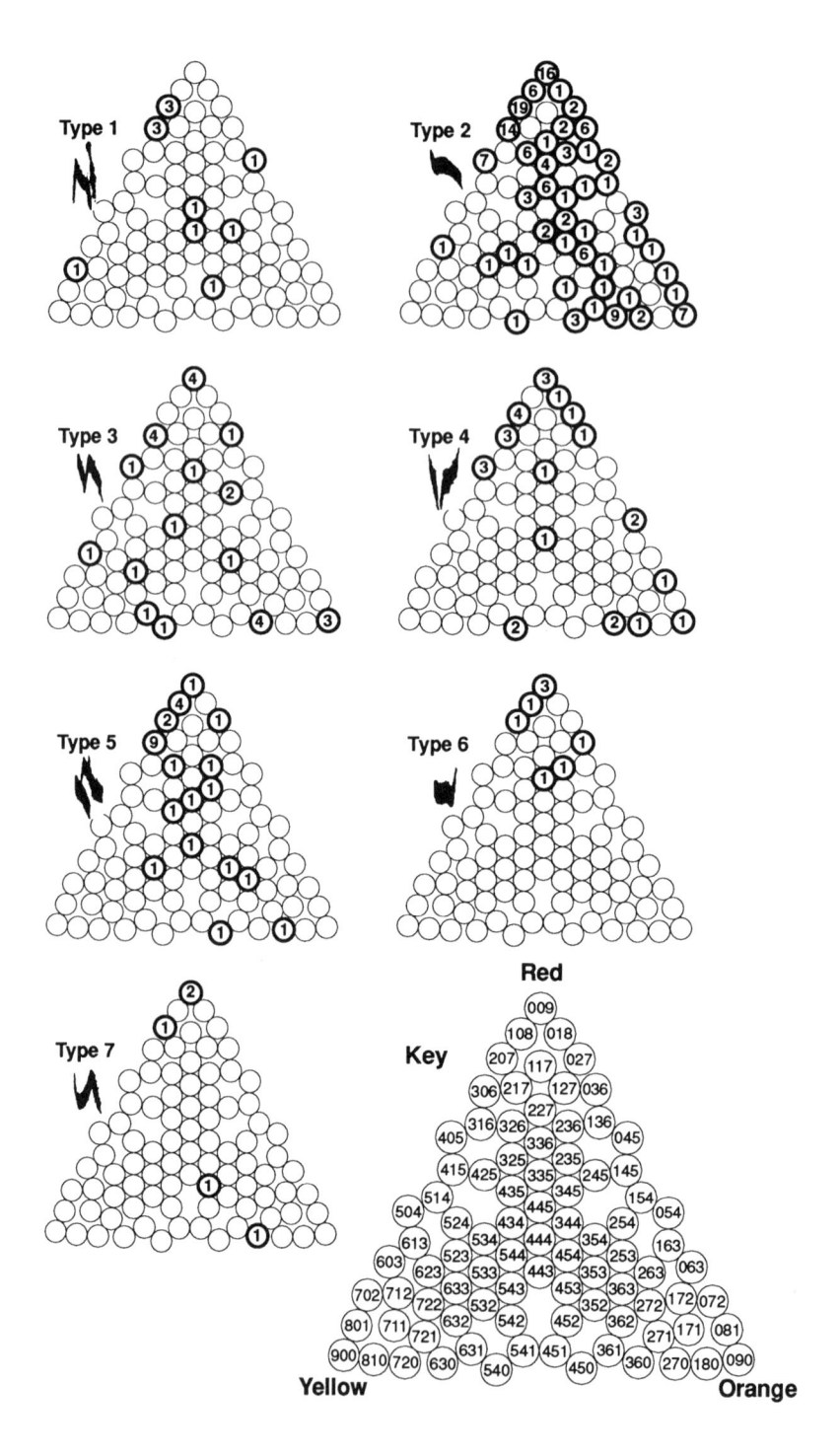

Figure 32. Distribution of color variants in adult male red crossbills of known vocalizations. Key to the diagrams is shown in the lower right; numbers in the circles of the key represent color codes (see Table 1). For the seven groups of crossbills, numbers in the circles give the count of specimens with the corresponding color code.

males molted wing and tail feathers and became more red in a "second immature" plumage. Tordoff (1954b) found a number of inconsistencies with Jollie's work, especially with the point regarding a yellow first immature plumage. Phillips' (1977) conclusions were similar to Ticehurst's (1915), in that he suggested that the most effective way of separating age classes of males was not by body plumage color but by the color of feather edges.

In this analysis, possible year-class differences were ignored; all birds with no traces of the streaked juvenal plumage were used, even if skull pneumatization was incomplete. Samples of the different units of analysis were not intentionally biased with regard to year-class distribution. Many partially streaked birds had new and emergent red feathers, therefore Jollie's (1953) conclusions were disregarded in favor of those of Ticehurst (1915), Tordoff (1952, 1954b), and Phillips (1977).

Distributions of color scores for the seven vocally defined groups of crossbills (Fig. 32) showed little difference in relative amounts of red or yellow in male plumages. All samples except Type 6 had large frequencies of males that were mostly orange or yellow. Males that had 50% or more "red" contour feathers (plumage scores with the third digit 5 or greater) comprised highly similar fractions of each sample: 67% in Type 1, 68% in Type 2, 54% in Type 3, 63% in Type 4, 73% in Type 5, 100% in Type 6, and 60% in Type 7. These results show little evidence that the amount of red or yellow in the plumage is related to variation in vocalizations. Griscom (1937) reported a greater frequency of "xanthochroistic" males among the smallest North American red crossbills, which is consistent with the slightly greater proportion of yellowish or orangish males in Type 3. Large frequencies of males in all groups were mostly red with a few yellow feathers around the face and particularly the throat region. It is presently not understood whether yellow face and throat feathers are the result of a prenuptial molt or casual losses (perhaps due to pine tar) with replacement. Color in males appears to vary in a way similar to the closely related house finch (*Carpodacus mexicanus*), in which coloration is related to diet (Brush and Power 1976, Hill 1992). In the males kept captive for this study, all molted into yellow in the first year.

One aspect of color that appeared to vary and was not studied was the amount of color saturation in contour feathers. It appeared that some groups, especially Type 3 (the smallest birds) and Type 6 (the largest), had deeper and more saturated colors than the others. Type 2 birds, in particular, were qualitatively paler than other groups. This variation may relate, at least in part, to the extent of deposition of pigment granules throughout the barbs and barbules of the feathers, which Gill (1990:75) has shown is the reason male white-winged crossbills (*L. leucoptera*) are more pinkish than male red crossbills.

ECOLOGICAL VARIATION

Sympatric congeneric birds should have ecological differences that allow sharing of the available habitat (e.g., Dilger 1956, Lack 1971), although the precise levels of morphological differentiation necessary for coexistence of ecological competitors have been debated (Hutchinson 1959, Schoener 1965, MacArthur and Levins 1967, Løvtrup et al. 1974, Horn and May 1977, James 1982, Schluter and Grant 1984, Pulliam 1985). Forms of North American red crossbills have varying degrees of bill size differentiation, therefore the complex may be appropriate for studying this problem.

Ecological observations. Many specimens in this study were kept captive for extended periods, and it was not possible to examine stomach contents of all birds; however, the major food sources of crossbills were usually obvious at each locality (see Groth 1990 for details). The only forms seen successfully extracting seeds from robust cones, including those of ponderosa pine (*Pinus ponderosa*), Apache pine (*P. englemannii*), and table-mountain pine (*P. pungens*), were Types 2 and 6, which have the largest bills. Although Type 2 and Type 5 crossbills have bills nearly identical in size, Type 5 birds were not seen using cones of ponderosa pine, even though Type 5 birds range extensively over western North America, where ponderosa pine is common. In contrast, Type 2 birds were commonly found in association with ponderosa pine and used this conifer as a food source while breeding. Types 2, 5, and 7 (which have similar bill sizes) were all found in association with lodgepole pine (*P. contorta*), and all three forms were found breeding sympatrically in lodgepole pine forests in the eastern Cascades of Oregon. In the lodgepole pine forests of California, Type 5 and Type 7 birds were observed far less commonly than Type 2. Type 2 birds have also bred successfully in eastern North America in association with other pines (see Groth 1988, 1990). The observation that crossbill forms with similar bill sizes differ in conifer usage suggests that factors other than bill size, such as cultural differences related to imprinting on different habitats (Immelmann 1975), may play roles in ecological segregation.

Small-billed forms, including Type 1, Type 3, and Type 4, were usually not associated with robust conifer cones. The ecology of the small-billed Type 3 form is poorly known. There are indications that tiny crossbills use hemlock (*Tsuga*) extensively (pers. obs.; T.

Hahn and C. Benkman, pers. comm.; see also Payne 1987). Type 4 birds showed strong association with Douglas fir (*Pseudotsuga menziesei*) in western North America, but the ecological preferences of this form in the east are not known. If Type 4 corresponds to "*neogaea*," it may be that eastern white pine (*Pinus strobus*) is the principal resource of this form in the east (see Dickerman 1987). Type 1 crossbills were strongly associated with eastern white pine in the southern Appalachians, and they have been found breeding in western hemlock (*Tsuga heterophylla*) forests on the Olympic peninsula of Washington (pers. obs.; T. Hahn, pers. comm.).

Although different forms of crossbill showed differences in conifer usage, considerable overlap was observed. Two or more forms occurring together at a site were often observed using the same conifers, while in other instances sympatric forms showed different conifer usage. Furthermore, many crossbills, including those in breeding areas, had crops filled with arthropods (see Groth 1990).

The relationship between bill size and cone size. Previous discussions of the evolution of crossbills have assumed that bill and body size differences are adaptive (Lack 1944a, 1944b, Knox 1975, 1990; Nethersole-Thompson 1975; Benkman 1989), and the selective force presumed in these discussions has been variation in cone morphology. For example, Benkman (1989) proposed adaptive explanations for bill morphology of a number of island crossbills, suggesting that bill forms are matched to the conifers on which the birds feed. This assumes that the forms are optimized for particular conifers and are not at intermediate stages of adaptive evolution. Alternative hypotheses are that the morphological differences among crossbill forms are the result of genetic drift or non-adaptive processes, and/or that different forms fly (by virtue of their high vagility and nomadic lifestyle) to areas containing conifers that their bills are best able to exploit.

Whereas there has been a suggestion that red crossbills rotate their use of conifers depending on cone ripening phenology (Benkman 1987), it would be useful to integrate information on the vocal and morphological identity of the birds observed. What appears to be conifer switching at a site might not involve the same set of individuals or even the same form of crossbill, and instead might relate to movements of different crossbill forms.

ALLOZYME VARIATION

Allozymes in the genus *Loxia* have been studied previously with regard to generic and family level systematics. Four specimens of *Loxia* were compared to other cardueline finches by Marten and Johnson (1986), showing that crossbills are genetically closer to *Carduelis* than to other North American genera. A specimen of *L. curvirostra* was used as a cardueline outgroup in N. K. Johnson et al.'s (1989) study of allozyme relationships of the Hawaiian honeycreepers. Other than these studies, the patterns of genetic variation within and among populations of the two North American crossbills, *L. curvirostra* and *L. leucoptera,* have not received attention.

Allozymes were studied to determine if the *curvirostra* complex is divisible into genetically differentiated units consistent with differences in vocalizations. Data from all *curvirostra* were first pooled to determine if the sample showed conformity to Hardy-Weinberg equilibrium expectations. As part of this first analysis, *curvirostra* was compared to *leucoptera*. The second analysis divided *curvirostra* into vocal groups.

GENERAL PATTERNS OF ALLOZYME VARIATION

All individuals were scored for 35 presumptive genetic loci. Of the loci scored for both *curvirostra* and *leucoptera,* 17 (EST-1, ALB-1, ALB-4, HB, GDA, LDH-1, LDH-2, CK-1, ICD-2, SDH, GLUD, EAP, GPT, MDH-1, MDH-2, ALD, and SOD-2) were monomorphic and fixed for the same allele in all individuals; the remaining 18 loci showed at least a single heterozygote (Appendix F). Of the 44 two-banded (heterozygous) individuals at monomeric ACON-1, all were males, which is in agreement with previous studies in birds showing a sex-linked mode of inheritance for this locus (Baverstock et al. 1982). For ACON-1, only results for males (the homogametic sex) were used in further computations of allelic frequencies, F-statistics, and genetic distances.

COMPARISON OF *L. CURVIROSTRA* AND *L. LEUCOPTERA*

Allelic frequencies and genetic distance. The *leucoptera* sample showed an unusually high mean heterozygosity value (across loci) of 12.7%, which is about twice that for birds in general (see Barrowclough et al. 1985) and also higher than values for other cardueline finches (Marten and Johnson 1986). The sample of *curvirostra* showed an average observed heterozygosity across loci of 7.0%, which is also higher than has generally been found in birds.

Populations of *curvirostra* and the sample of *leucoptera* had highly similar gene frequencies (see Appendix F). Nei's (1978) genetic distance between *curvirostra* and *leucoptera* was 0.019. At all loci, the most common allele in *curvirostra* populations was also the most common allele in *leucoptera*, with the single exception of ICD-1, in which the two most common alleles were in different proportions. No loci that were polymorphic in *leucoptera* were not also polymorphic in *curvirostra*. Marten and Johnson (1986) reported a fixed difference at the ADA locus between one specimen of *curvirostra* and three of *leucoptera*, but this could not be repeated after several tries with larger samples in which banding appeared identical. Of the 59 alleles found in *leucoptera*, only 5 were not also represented in the *curvirostra* sample. Frequencies of alleles unique to *leucoptera* were low, ranging from 0.031 to 0.094. A total of 99 alleles was detected in *curvirostra*, of which 45 were not found in *leucoptera*; the frequencies of these were also low, ranging from 0.001 (most occurrences) to 0.313 (LA-2^{110}). The mean frequency of alleles found in only one species (n = 50) was 0.0213. Considering the two species as "demes" and using the method of Slatkin (1985a) yielded an estimate of Nm (corrected for mean sample size of 288.5) of 1.41.

Intraspecific variation. If the pooled sample of 561 *L. curvirostra* included genetically distinct populations, there should have been a resulting deficiency of heterozygotes, termed the "Wahlund effect" (Wahlund 1928, Wallace 1981). Loci were equally divided between those with either excesses or deficiencies of heterozygotes, and all deviations were low. The mean F across polymorphic loci (equal weighting) was 0.018. The 5 loci with the highest heterozygosities (ICD-1, LA-1, LA-2, LGG, and 6-PGD) also had the highest F values, which ranged from 0.025 to 0.091, indicating moderate deficiencies of heterozygotes. However, chi-square test with pooling into three classes of genotypes per locus (Swofford and Selander 1981; see chapter above on Materials and Methods) showed no significant ($P < 0.05$) departures from Hardy-Weinberg expectations. These results do not falsify the hypothesis that North American *L. curvirostra* reflects a single Hardy-Weinberg population.

The sample of *leucoptera* stood in contrast to *curvirostra* in showing observed numbers of heterozygotes in excess of Hardy-Weinberg expectations (negative F values) at all 12 polymorphic loci. This result was unexpected because pooling of individuals from widely separated areas (see Materials and Methods) might have resulted in a Wahlund effect. Although these observations provide no evidence for genetic substructure in this species, low sample size does not allow further speculation on this anomalous result.

ALLOZYME COMPARISONS OF *L. CURVIROSTRA* VOCAL GROUPS

General patterns and levels of variability. The most common allele at each locus was the same for all seven *curvirostra* populations, with the single exception of a reversal in relative frequencies of the *100* and *111* alleles at LGG in the Type 1 population. Another shift was the relatively high frequency of 6-PGD^{87} in Type 6. Other fluctuations were slight. The forms had highly similar mean heterozygosities across loci, with values ranging from 6.3% to 8.2%. Mean numbers of alleles per locus (Table 13) ranged from 1.4 (Type 6) to 2.5 (Type 2), and frequencies of polymorphic loci ranged from 31.4% (Type 7) to 51.4% (Type 2). Variation in the latter two parameters was clearly correlated with sample size and probably not indicative of true differences in within-population genetic variability.

Genetic distances and branching diagrams. Genetic distances among forms of *curvirostra* were low (Table 14). Nei's (1978) distances ranged from 0.000 to 0.006 (n = 21, mean = 0.0010, SE = 0.0003), and Rogers' (1972) distances ranged from 0.008 to 0.040 (n = 21, mean = 0.0215, SE = 0.0018). The branching pattern in a distance Wagner tree (Fig. 33A, rooted at *leucoptera*) shows linkage between the Type 6 and Type 5 populations, although the branch length between them is long. Type 5 and Type 6 are linked to Type 2, and this cluster joins first with Type 3, Type 4, and Type 7. Type 1 connects last before the *curvirostra* cluster is linked to *leucoptera*. The conformation in the 3-D multidimensional scaling plot (Fig. 33B) was reached after 17 iterations when a stress of less than 0.001 was reached. This plot shows that the most genetically divergent forms of *curvirostra* were Type 1 and Type 6.

Population subdivision and gene flow. Allozyme differences among populations were not a significant source of genic variation within *curvirostra*. Nei's (1977) F_{ST} ("G_{ST}") values ranged from 0.002 to 0.051 across polymorphic loci. Three loci (LGG, 6-PGD, and ACP) exhibited significant (using the formula of Workman and Niswander 1970) between-population heterogeneity in Wright's (1978) F_{ST} (see Groth 1990), and the mean value of Wright's F_{ST} across polymorphic loci was 0.0056 (SE = 0.0029), which indicates that less than 1% of allelic variance was contributed by differences among populations.

The inbreeding coefficients (F_{IS}) were negative (indicating heterozygote excess) in 11 of 18 loci, with a mean across polymorphic loci of 0.0154. F_{IS} values are weighted averages across populations and thus differ from the F values in Table 13, in which all individuals were pooled. The strongest deficiencies of heterozygotes within populations were at GPI (F_{IS} = 0.247) and 6-PGD (F_{IS} = 0.165). The case of one Type 6 bird homozygous for the GPI^{152} allele, with the other 14 Type 6 individuals homozygous for GPI^{100}, was unusual.

Three populations (Type 5, Type 6, and Type 7) did not show any unique alleles. Type 1 had 1, Type 2 had 15, Type 3 had 4, and Type 4 had 1, for a total of 21 unique alleles over all call types; frequencies ranged from 0.002 to 0.018. The number of unique alleles within populations was apparently dependent on sample size. All unique alleles were from single occurrences in individuals, with the exceptions of PAP^{105} and GOT-2^{0}

Table 13. Number of Alleles per Locus, Expected and Observed Numbers of Heterozygotes, and Fixation Indices (Inbreeding Coefficients, F)[a] for 18 Polymorphic Loci in *L. curvirostra* and *L. leucoptera*

Locus	*L. curvirostra* (n = 561) Number of alleles	Number of heterozygotes expected	observed	F	*L. leucoptera* (n = 16) Number of alleles	Number of heterozygotes expected	observed	F
ICD-1	4	204.17	199	0.025	4	9.59	10	-0.042
MPI	5	55.45	55	-0.010	1	—	—	—
α-GPD	5	53.99	54	0.000	3	2.78	3	-0.079
NP	6	36.22	35	0.034	3	3.59	4	-0.113
PGM	5	17.80	18	-0.011	2	0.97	1	-0.032
GPI	7	35.19	32	0.091	1	—	—	—
CK-2	3	8.95	9	-0.006	1	—	—	—
EST-4	2	7.94	8	-0.007	2	2.72	3	-0.103
LA-1	8	159.52	148	0.072	2	1.88	2	-0.067
LA-2	9	253.39	235	0.073	3	8.88	13	-0.465
LGG	5	297.27	287	0.035	4	7.19	8	-0.113
PAP	3	2.99	3	-0.002	1	—	—	—
GOT-1	4	7.96	8	-0.005	5	9.75	12	-0.231
GOT-2	3	8.94	9	-0.007	1	—	—	—
ACON-1[b]	3	44.03	44	0.001	1	—	—	—
6-PGD	3	124.15	117	0.058	2	4.22	5	-0.185
SOD-1	4	9.94	10	-0.006	4	7.09	8	-0.128
ACP	3	66.57	67	-0.006	2	1.88	2	-0.067

a. (Expected minus observed number of heterozygotes) divided by the expected number of heterozygotes.
b. Males only (n = 325 for *L. curvirostra*).

(two individuals each) and NP[89] (nine individuals) all in Type 2. The mean frequency of unique alleles in populations was 0.0050 (SE = 0.0010). The value for Nm using Slatkin's (1985a) formula was 89.7 (corrected using an average sample size of 80), which is a value suggesting high gene flow.

Table 14. Genetic Distance Estimates[a] Among 7 Forms of *L. curvirostra* and One Sample of *L. leucoptera*

	Type 1	Type 2	Type 3	Type 4	Type 5	Type 6	Type 7	*leucoptera*
Type 1	—	0.001	0.001	0.001	0.002	0.006	0.000	0.019
Type 2	0.022	—	0.000	0.000	0.000	0.001	0.000	0.019
Type 3	0.023	0.014	—	0.000	0.001	0.003	0.000	0.020
Type 4	0.022	0.013	0.008	—	0.001	0.003	0.000	0.021
Type 5	0.023	0.010	0.018	0.016	—	0.000	0.000	0.019
Type 6	0.040	0.025	0.033	0.033	0.020	—	0.001	0.019
Type 7	0.024	0.018	0.018	0.018	0.020	0.034	—	0.019
leucoptera	0.065	0.066	0.070	0.070	0.068	0.074	0.072	—

a. Nei's (1978) distances in upper matrix; Rogers' (1972) distances in lower matrix.

DISCUSSION

Although *L. curvirostra* and *L. leucoptera* differ in plumage coloration, morphology, and vocalizations, they are only weakly differentiated at allozyme loci. Average Nei's *D* among congeneric bird species has been estimated at 0.100 (Barrowclough et al. 1981, Avise et al. 1982), a value much higher than that between *curvirostra* and *leucoptera* (*D* = 0.019). There is no evidence that the low genetic distance between the crossbill species was due to hybridization and genetic introgression. Although the two species overlap broadly in geographic ranges and are highly similar in general life histories, no natural hybrids between them are known. This information appears to conflict with the *Nm* values between the two species (as "demes" using Slatkin's [1985a] method) which was approximately 1.4, a value suggesting high gene flow. One explanation is that the total deme (species) sizes (*N*) is in the tens or hundreds of thousands, and the actual migration rate (*m*) is therefore minute.

Gutiérrez et al. (1983) and Marten and Johnson (1986) have proposed calibrations using Nei's *D* values for estimating timing of phyletic divergence. The estimates for the two studies were at 26.3 and 19.7 million years before present, respectively, for every one unit of Nei's (1978) *D*. Using these calibrations, the estimated separation of *curvirostra* and *leucoptera* is less than one million years ago. These estimates should be considered only very rough because of questions regarding rate variation in different taxa (e.g., Ayala 1986, Vawter and Brown 1986) and the fact that fossils used for the calibrations are of questionable age (Johnson and Marten 1986). Furthermore, genetic distance estimates themselves are highly subject to error because of biased sampling of loci.

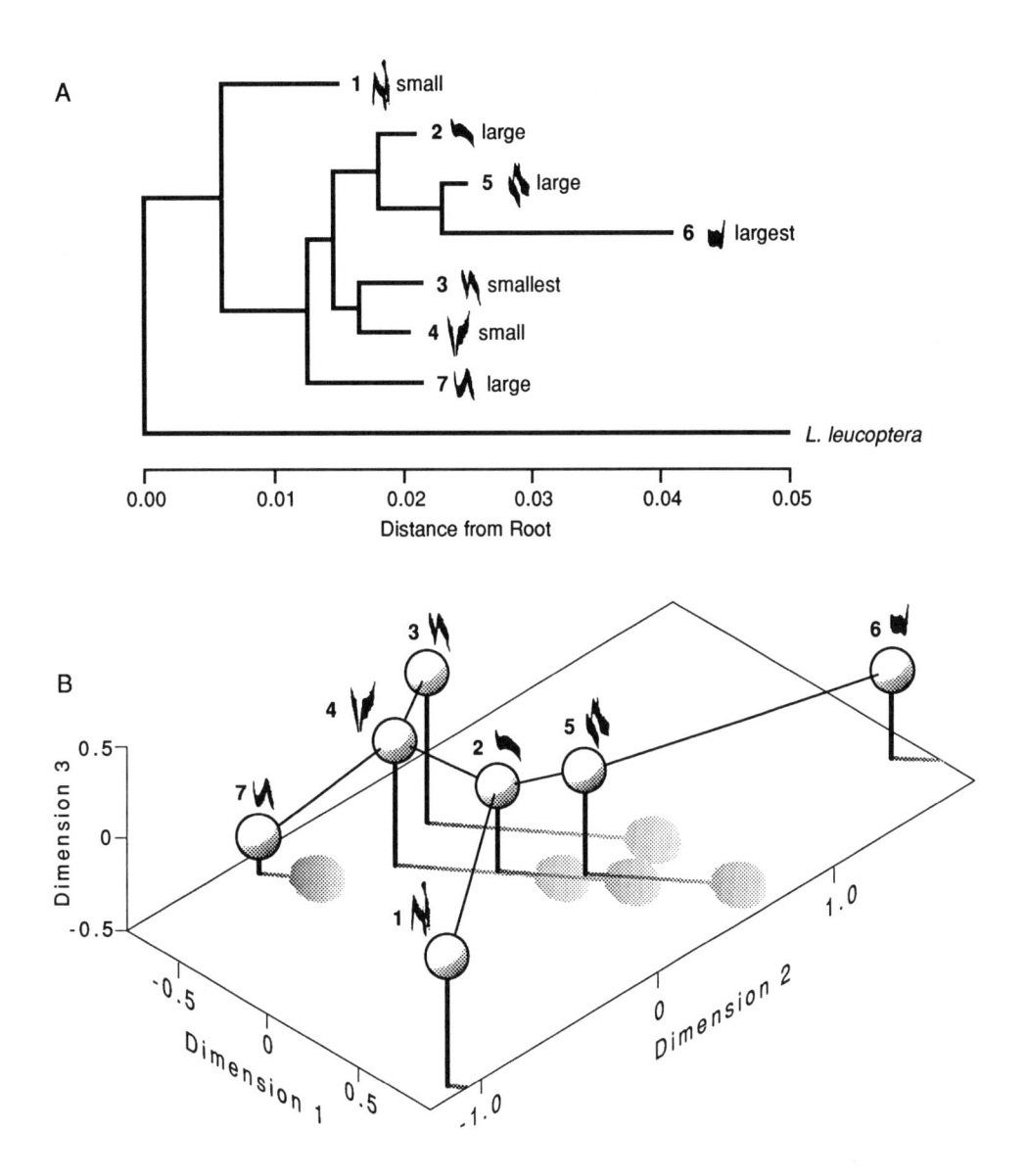

Figure 33. Phenograms portraying Rogers' (1972) genetic distances. (A) Distance Wagner phenogram for 7 groups of red crossbills and one sample of *L. leucoptera*, rooted at *L. leucoptera*. (B) Multidimensional scaling plot with minimum spanning tree; r_{cc} of 0.987 indicates good fit to original distances.

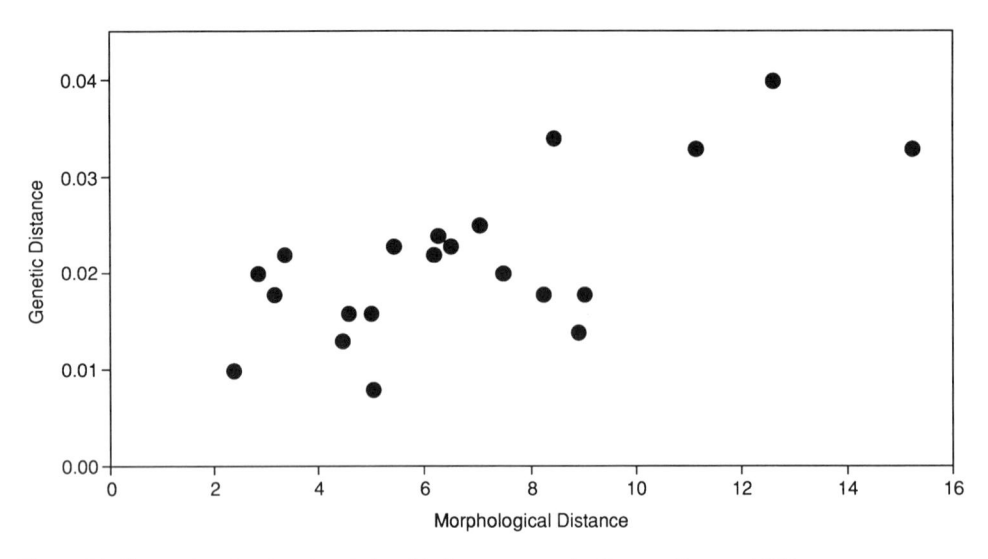

Figure 34. Scatterplot showing the relationship between genetic distance (Rogers 1972) and morphological distance (Mahalanobis' D^2 values, based on 32 morphometric characters in canonical discriminant analysis of adult males, Table 12) in pairwise comparisons among 7 vocally defined groups of red crossbills.

If vocally defined groups of red crossbills are separate lineages, calibrated estimates of divergence times would fall within the last 100,000 years (Marten and Johnson 1986). Although the current hypothesis is that these groups represent reproductively isolated species, their level of differentiation was exceeded by several conspecific avian populations (e.g., M. C. Baker et al. 1982, A. J. Baker and Moeed 1987, Johnson and Marten 1988, A. J. Baker et al. 1990). However, low allozyme distances do not falsify the hypothesis that different call types are species, because distances among other recognized avian species are similar to those found among crossbills (e.g., Avise et al. 1980a, 1980b; Yang and Patton 1981; Johnson and Zink 1983; Zink and Johnson 1984; Braun and Robbins 1986; Avise and Zink 1988). Avian populations and sibling species with low allozyme distances may be genetically different in their mitochondrial DNAs (e.g., Avise and Zink 1988, Zink 1991) as well as different in genes for the morphological or behavioral traits that provided the original bases for taxonomic recognition of these forms.

Low F_{ST} values and high Nm values among call types suggest little genetic differentiation and high gene flow. The Nm value of 89.7 estimated among call types using Slatkin's (1985a) method exceeds previous estimates for avian conspecific populations (Rockwell and Barrowclough 1986, Zink et al. 1987, A. J. Baker et al. 1990). As pointed out by Zink and Remsen (1986), computation of Nm values across full species considered as "demes" can give values even higher than some estimates within populations of the same species. Furthermore, Slatkin (1985b) cautioned that Nm estimates are only valid if populations have reached genetic equilibrium. This may not be the case in crossbills, and other passerines, if populations diverged recently and population sizes are large.

It is evident that the "small" crossbill forms (Type 1, Type 3, and Type 4) do not form a cluster based on allozymes (Fig. 33). However, that the population most genetically similar to the largest form (Type 6) was large (Type 5) indicates some concordance between character sets. There was an overall correspondence between genetic distances and morphological distances among red crossbill forms (Fig. 34), and the correspondence between matrices using Mantel's test ($t = 2.114$, $0.05 > P > 0.02$) was significant. However, if Type 6 (the most morphologically and genetically divergent form) is deleted from both matrices, the trend for a positive relationship disappears and the matrices show no significant correspondence (Mantel's test, $t = 0.469$, ns).

Structures of vocalizations varied in their agreement with allozyme results. One example of concordance was the similarity of the alarm and excitement calls of the Type 5 and Type 6 populations (see Fig. 13), all of which were short in duration and contained closely overlying frequency bands. An example of discordance was that Type 2 and Type 5 have divergent alarm and excitement calls (as well as distinctive flight calls), yet they were highly similar in allozymes. Type 3 and Type 4 were nearly identical in allozymes, yet their call repertoires were highly distinctive in all respects.

DISCUSSION AND CONCLUSIONS

LEVEL OF DIFFERENTIATION

Previous reviewers (e.g., Griscom 1937, Monson and Phillips 1981, Dickerman 1986a) treated crossbills differently from other organisms in discussing distinct forms as subspecies with geographic overlap. Monson and Phillips (1981) called their four size classes "semi-species"; however, the sympatry of crossbill forms precludes the terms population, subspecies, and semispecies, because these terms represent geographically separated groups (Mayr 1963, Futuyma 1986). The view favored here is that crossbill forms are sibling species (morphologically similar or identical populations that are reproductively isolated; Mayr 1963), meaning that crossbills need not be exceptions to traditional ontology used in systematic biology. Vocally and morphologically differentiated forms of crossbills conform to a phylogenetic species concept, in which a species is "the smallest diagnosable cluster of individual organisms within which there is a parental pattern of ancestry and descent" (Cracraft 1983), but crossbill forms are unusual in this context because the characters used for their diagnosis were primarily behavioral. Furthermore, local aggregations of crossbills that settle and breed on ephemeral conifer cone crops should not be considered "populations," because (1) the aggregate may contain a mixture of sibling species, and (2) the birds will probably disperse after the food supply runs out and not reassemble in any consistent fashion.

An argument in opposition to the species level in crossbills takes the position that the component forms are simply not different enough morphologically to deserve full species status. For example, in a study of species limits in redpolls (*Carduelis flammea* and *C. hornemanni*), Troy (1985) expressed the view that phenetic overlap was evidence for conspecificity. However, in other cases in which the species level is undisputed, such as in the brown towhee complex (*Pipilo*; Zink 1988), "good" species overlapped in multivariate morphospace in a way similar to the forms of crossbills.

The suggestion by several authors (Meinertzhagen 1934, Dement'ev et al. 1965) that the parrot and common crossbills of Eurasia are morphs was discussed and found unlikely by Knox (1990), who considered the two forms different species. "Morphs" within populations should interbreed; and as a result of Mendelian inheritance, groups of siblings will often be of mixed morphotype (Huxley 1955). Because flight call structures in crossbills are learned, the polymorphism hypothesis also implies that there should be some instances of vocally defined groups of crossbills containing more than one morphotype, but this was not observed.

With the high mobility and frequent mixing of crossbill forms, there is ample opportunity for mate exchange among them, yet interbreeding (if any) has not homogenized morphology in the complex. Furthermore, mated pairs contain members of the same morphotype in areas where different forms are sympatric (Groth 1993). It is presently not understood how these forms avoid interbreeding in the wild. Vocal differences among crossbill forms are discrete enough to function as "specific mate recognition systems" in the sense of Paterson's (1985) "recognition" concept of species. Crossbills might also visually assess the morphology (bill size) of potential mates, as has been demonstrated in Darwin's finches (Ratcliffe and Grant 1985), but this would not be effective in preventing interbreeding between some pairs of nearly identical sympatric crossbill forms such as Type 2 and Type 5.

SPECIATION IN NORTH AMERICAN RED CROSSBILLS

Clusters of individual crossbills of similar morphology and vocalizations appear to be the most appropriate units for phylogenetic analysis (Cracraft 1989) within this complex. This study did not construct a phylogeny because (1) there are numerous problems in using vocalizations in phylogenetic analysis (Payne 1986); (2) the morphological characters measured here were not discrete; and (3) the allozyme differences among forms were too small. Lacking phylogenetic analysis, it is presently not possible to definitively address patterns or processes of speciation in North American red crossbills.

It is not known if the North American complex is a monophyletic assemblage. Ridgway (1885a) was among the first to notice the strong morphological similarity between east Asian red crossbills ("*japonica*") and many specimens from western North America ("*bendirei*"). Griscom (1937) later proposed that the red crossbill lineage originated somewhere in Asia, with subsequent migration of some forms over the Bering land bridge to the New World. Griscom (1937:207) wrote, "the very close resemblance of *benti* to *japonica* and *minor* to *himalayensis* suggests the necessity of postulating arrival of at least *three* different stocks from the Old World, unless we choose to assume an amazing parallelism in evolution or accidental convergence of characters, a most unsound proposition biologically." Thus Griscom favored the idea that North American crossbills are polyphyletic. Red crossbills have probably crossed the Bering Strait more than once in the present century (e.g., Stepanyan 1979), yet most such movements must go undetected. This suggests that some Old World populations could be derivatives from New World sources.

Pollen and macrofossils show little evidence for past discontinuities in conifer distributions that might have separated continental North American crossbill populations. Much of what is now the Great Plains was once covered by pine and spruce woodland or parkland (P. V. Wells 1970, Delcourt and Delcourt 1987), which could have been used as an east–west dispersal corridor by crossbills. Alternatively, different species of conifers may have been geographically separated in the past, and the different bill morphologies of crossbills could be the result of adaptation to region-specific conifers. For example, forests dominated by eastern white pine (*Pinus strobus*) and eastern hemlock (*Tsuga canadensis*) were confined to a relatively small area along the middle Atlantic coast of the United States during the last glacial maximum (M. B. Davis 1983), and this may have served as a pocket of diversification for a small-billed form of crossbill adapted for small and soft cones.

Several forms of crossbills on islands, such as *scotica* on Scotland, *percna* on Newfoundland, and *luzonensis* in the Philippines, are morphologically distinctive. Rapid genetic diversification on islands would be promoted by founder events and small population sizes (Boag and Grant 1981, Barton and Charlesworth 1984, Carson and Templeton 1984, Baker and Moeed 1987). New forms could then move off islands and behave as distinct species when colonizing continental areas already inhabited by other distinctive crossbill forms.

A plausible mechanism for crossbill speciation is that populations initially became subdivided geographically, under a variety of circumstances, for periods of time long enough for rapid cultural evolution (Theilcke 1973, Bonner 1980, Mundinger 1980, Ince et al. 1980, Payne et al. 1981, Cavalli-Sforza et al. 1982) in vocalizations and possibly also in habitat preference (Immelmann 1975). After secondary contact and sympatry, populations would remain as separate reproductive communities by virtue of their behavioral differences alone. Genetic and morphological diversification would later accumulate through stochastic factors (genetic drift) and/or natural selection (microevolution). Further divergence in vocal characters could emerge through character displacement (E. H. Miller 1982) and/or ecological sources of selection (Morton 1975, Bowman 1979).

It is interesting to ask why there are so many kinds of crossbills. Lauder (1981) has stressed that "key innovations" typify adaptive radiations, and the crossed bill of *Loxia* appears to be such an innovation. The variety of "niches" presented by the diversity of conifer cones may be a zone in which adaptive radiation (in bill size) has been possible. Another contributing factor in crossbill diversification may relate to the one discussed by Wyles et al. (1983), in which large brain size and learning ability has driven rapid morphological evolution in birds; in this model, changes in culture may lead to changes in habitat usage which are then followed by changes in the forces of natural selection. Whereas this process deals with anagenesis (changes within lineages), social behavior may also promote cladogenesis (speciation) through "behavioral drive" (Larson et al. 1984) favoring establishment of new populations and evolution of mate recognition signals. These ideas are similar to those discussed by Raikow (1986), in which the evolutionary explosion of passerines is correlated to complex syrinx structure and vocal learning ability.

Endler (1977) outlined a mechanism by which geographically contiguous (parapatric) populations may speciate. In this model, different frequencies of co-adapted gene complexes initially vary clinally over an environmental gradient, and under certain circumstances assortative mating would develop given sufficient selection against heterozygotes. One criterion for the model is that "only a *very small fraction* of the members of a given group are within 'cruising range' of the others" (p. 15, emphasis his). The enormously high "cruising" ability of nomadic continental populations of crossbills makes them unlikely candidates for the parapatric model.

The observed broad sympatry, high mobility, and nomadism of morphologically similar forms of crossbills have been stumbling blocks for allopatric speciation models in this group. Although most proposed examples of sympatric speciation are from work on insects (Bush 1975, review in Tauber and Tauber 1989) and fish (e.g., several authors in Echelle and Kornfield 1984), few ornithologists have centered their observations around sympatric speciation models (for exceptions, see B. R. Grant and P. R. Grant 1979; P. R. Grant and B. R. Grant 1989; T. B. Smith 1990a, 1990b). Models of sympatric divergence generally involve separation of lineages within a single area, based on habitat assortment or niche diversification, followed by assortative mating (Maynard Smith 1966, Bush 1975, Tauber and Tauber 1977, Felsenstein 1981, Rice 1984, Diehl and Bush 1989). Many forests contain more than a single species of conifer, reflecting different ecological zones for crossbills to occupy in sympatry. A beginning step in these models is the occurrence of discrete morphs within populations. At this hypothetical stage, populations would exhibit polymorphism without advancement to positive assortative mating, such as in *Pyrenestes* finches (T. B. Smith 1987), yet if this were the predominant speciation mode in crossbills, an expectation would be that extant groups would exhibit polymorphism. Another sympatric speciation mechanism outlined by Nicolai (1964) and elaborated upon by Payne (1973) for brood parasitic finches (*Vidua*) was that vocal learning and imprinting after rare events of switching of host species (song models) could give rise to different reproductive communities within a single generation. Differentiation in *Loxia* resembles that in *Vidua* in that both contain sibling species distinguishable in large part on the basis of learned vocal characters. Crossbills, however, are not brood parasitic, but alternative conditions which might produce drastic vocal shifts in crossbills are presently unknown.

NOMENCLATURE

As species, crossbill forms will require both scientific binomial names (International Trust for Zoological Nomenclature 1985) and common names. Nomenclature within *Loxia* has historically been among the most contentious for any group of birds, and with the new interpretation that the complex is a group of sibling species, the existing system of names will require alteration. However, this task will be a difficult one. Because of morphological overlap among groups, many type specimens cannot be assigned with certainty. A novel approach might include comparisons of DNA extracted from the old type specimens (Pääbo 1989, Thomas et al. 1990, Diamond 1990) to DNA from birds of known vocalizations. The section below indicates some tentative taxonomic conclusions and several unsolved problems regarding the nomenclature in North American red crossbills.

The smallest crossbills all fall into a single distinctive group (Type 3) which coincides morphologically with "Class I" of Monson and Phillips (1981). After measurement of type specimens and other series of crossbills in this size range, Payne (1987) suggested that *minor, sitkensis,* and *reai* are synonyms, with *minor* taking priority. This decision appears justified in the light of vocal information.

Two slightly larger forms, Type 1 and Type 4, match the measurements of Monson and Phillips' (1981) "Class II." This suggests that two names should apply to crossbills in this size range, but the question of which type specimens belong to which vocally defined form remains problematic. Monson and Phillips (1981) listed *minor, bendirei, pusilla,* and *neogaea* as names that have historically represented birds in this size class, then added *vividior* for a southern Rocky Mountain form. Monson and Phillips (1981), Payne (1987), and Groth (1988) agreed that the type specimen of *minor* is too small to represent birds of this size. I further showed (Groth 1988) that the type of *pusilla* is too large to represent Type 1 birds; therefore, this specimen is also too large to represent Type 4. Payne (1987) showed that several specimens in the type series of *bendirei* from Fort Klamath, Oregon, were larger than the type of *pusilla*, hence many in the series were larger than Type 1 and Type 4 birds. Monson and Phillips (1981) considered *bendirei* to represent larger birds (their Class III), which agrees with the relatively high frequencies of Type 2 and Type 5 crossbills found on the eastern slope of the Cascade range (near Fort Klamath) in the present study. Specimens with tags pencil-marked "*neogaea*" by Allan R. Phillips, and the measurements given by Monson and Phillips (1981) for *vividior*, match Type 1 and Type 4 birds. The present problem is whether *vividior* and *neogaea* are synonyms. Type 4 crossbills range over the type localities for both names, suggesting that *neogaea* would be the appropriate name for this form, with *vividior* as a synonym. This would leave the Type 1 form without another available name.

Larger crossbills, including Type 2, Type 5, and Type 7, have measurements matching Monson and Phillips' (1981) "Class III." Measurements taken from the type specimen of *pusilla* (which probably came from Georgia) fell within the ranges of all three forms. The evidence that *pusilla* should represent Type 2, rather than Type 5 or Type 7, is circumstantial in that only Type 2 birds have been recorded in the southern Appalachians. The names *bendirei, benti,* and *grinnelli* have also been used to refer to crossbills in this general size range. Measurements of the type specimens (unpubl. data) place all three within the size ranges of Type 2, Type 5, and Type 7. The three names may all be synonyms of *pusilla* and all represent Type 2 birds. Type 2 was the common form in the Sierra Nevada of California where the type of *grinnelli* was taken, and Type 2 was also recorded in the Dakotas near the type locality of *benti*. Type 2, Type 5, and Type 7 have all been recorded in Klamath County, Oregon, near the type locality of *bendirei*, and all three forms have measurements considerably overlapping those cited in Payne (1987) for the *bendirei* series. Unless *bendirei* is found to be a synonym of the older name *pusilla*, it should become available for either Type 5 or Type 7.

Type 6 crossbills match Monson and Phillips' (1981) "Class IV." No names other than *stricklandi* have been applied to crossbills in this morphological class since the work of Ridgway (1885b). One reason for caution is that it is not known if these largest crossbills comprise a single form based on vocalizations.

Most crossbills collected in Central America are morphologically different from those in the present study. Although generally about the size of Type 2 birds, most have shorter wings, shorter tails, and longer bills. Howell's (1972) suggestion of two Central American forms should serve as a motive for further work on the vocalizations and morphology of crossbills in that region.

Many Newfoundland crossbills are morphologically different from those of any of the specimens investigated in the present study (Payne 1987; Groth, unpubl. data). These data, and the recordings of a distinctive call type ("Type 8"), suggest that a unique taxon exists on the island. The name for this form is most appropriately *percna* Bent 1912.

In summary, the name *minor* should probably be assigned to Type 3 and *stricklandi* to Type 6. The name *pusilla* shows the best fit to Type 2, and it is probable that *benti* and *grinnelli* are synonyms of *pusilla*. The name *bendirei* is either a synonym of *pusilla*, or it could be applied to Type 5 or Type 7. The name *neogaea* (=*vividior*?) belongs to either Type 1 or Type 4.

SUMMARY

For those who have considered it a single species, the red crossbill has presented the contradictory duality of nomadism with strong "subspecific" morphological variation. Results presented here offer a simpler explanation: *L. curvirostra* is a group of sibling species, and the potential for gene flow derived from high vagility has not homogenized morphology in the complex because the different forms are reproductively isolated.

Audiospectrography of crossbill vocalizations from widely separated regions of North America shows eight distinctive variants corresponding to morphologically distinctive forms. Each vocally defined group of crossbills has its own narrow range of morphological variation which is a fraction of the total variation in the complex. Crossbill forms are characterized by constancy in levels of within-group morphological variation and sexual dimorphism. Character wear contributes some within-form variation, especially in bill length, but age, as well as leg length differences associated with bill crossing direction, do not. Geographic variation in morphology or vocalizations within forms was not evident, even over thousands of kilometers.

Univariate and multivariate morphometric analyses show highly significant mean character differences among the different forms, and many pairs of forms are morphologically non-overlapping. Other forms are nearly identical in morphology. Overall morphological differentiation is greatest in bill size characters. Plumage color differences are slight; both red and yellow males are found in each form. Crossbills with different bill sizes are generally associated with different conifer species. Allozyme frequency differences among the forms are slight, as are those between *L. curvirostra* and *L. leucoptera,* suggesting either recency of divergence or ongoing gene flow. The hypothesis of recent divergence is currently favored because of the morphological and behavioral evidence for non-interbreeding.

The overall pattern in the complex showing smaller-bodied birds in the northern part of the range is explained not as true clinal variation, but instead as a reflection of different frequencies of sibling species over geography. Four of the forms are known to range from the western Rocky Mountains east to Appalachia and the Atlantic coastal region. At least six forms occur in the Pacific Northwest, where conifer diversity is great and crossbills are common members of the local avifauna.

Appendix A. List of Field Localities and Sample Sizes

Locality	Date	Males	Females	Juveniles[b]
Northwestern Region				
Willow River, 2 mi. S + 16 mi. E Prince George, elev. 2800 ft., British Columbia	1 August 1987	–	–	1
Bowron River, 21 mi. E + 3 mi. S Penny, elev. 2800 ft., British Columbia	3 August 1987	–	–	8
Thompson Plateau, 7 mi. E + 21 mi. S Clinton, elev. 2900 ft., British Columbia	4–10 August 1987	15	8	20
Sheep Creek, 2 mi. N + 1 mi. W Northport, elev. 1900 ft., Stevens Co., Washington	29–31 July 1986	3	–	14
North Snow Peak, 12 mi. S + 3 mi. W Clark Fork, elev. 4000 ft., Shoshore Co., Idaho	2 August 1986	1	–	–
Cherry Creek, 4 mi. W Easley Peak, elev. 7400 ft., Blaine Co., Idaho	4–6 August 1986	1	5	18
Oregon				
1/2 mi. S + 1/2 mi. E Sea Lion Point, elev. 650 ft., Lane Co.	24–25 July 1986	3	3	12
Baker Beach forest, 2 mi. S Sea Lion Point, sea level, Lane Co.	22 July 1988	1	2	–
Elder Mountain, 3 mi. E + 1 mi. N Takilma, elev. 3500 ft., Josephine Co.	19 June 1985	1	3	–
Base of Saddle Mountain, 12 mi. E + 2 mi. S Chiloquin, elev. 4700 ft., Klamath Co.	6–7 June 1985	9	2	–
(same as last entry)	18 April 1988	1	–	–
5 mi. SE Mt. Thielsen, elev. 4700 - 6100 ft., Klamath Co.	11–13 August 1986	3	3	8
Odell Creek, 7 mi. NE Crescent Lake, elev. 4400 ft., Klamath Co.	24–26 July 1988	14	3	–
Willow Creek, 10 mi. SE Lakeview, elev. 6000 ft., Lake Co.	27 July 1988	2	1	–
Northern and coastal California				
Lake Mt., 7 mi. NW Hyampom, elev. 3100 ft., Humboldt Co.	31 May 1985	5	3	–
Copper City, 7 mi. NW Alder Springs, elev. 5900 ft., Glenn Co.	25 June 1985	1	1	–
University of California Field Station, elev. 1300 ft., Alameda Co.	25 August–25 November 1984	20	12	2
Inverness Ridge, 3 mi. NW Inverness, elev. 300 ft., Marin Co.	27 December 1984	7	2	–
Skegg's Point, 4 mi. W Woodside, elev. 1800 ft., San Mateo Co.	28 December 1984	1	–	–
Bolinas Ridge, 4 mi. E Bolinas, elev. 1600 ft., Marin Co.	13 March 1985	7	4	–

Appendix A (continued)

Locality	Date	Birds Examined[a] Males	Females	Juveniles[b]
Sierra Nevada of California				
Sagehen Creek Field Station, 3 mi. NW Hobart Mills, elev. 6400 ft., Nevada Co.	13–14 October 1984	22	14	–
(same as last entry)	26–27 October 1985	2	6	5
(same as last entry)	3 May 1986	–	2	–
(same as last entry)	11 June 1987	2	1	–
Barnes Flat, 11 mi. N + 4 mi. W Westwood, elev. 5500 ft., Lassen Co.	31 August 1987	3	–	1
Prosser Creek, 1 mi. W Hobart Mills, elev. 6000 ft., Nevada Co.	29 June–1 July 1988	11	8	–
Arizona				
Tornado Clearing, 17 mi. S Jacob Lake, elev. 9100 ft., Kaibab Plateau, Coconino Co.	11–12 June 1986	2	1	–
Strayhorse Divide, 11 mi. S + 1 mi. W Hannagan Meadow, elev. 7100 ft., Greenlee Co.	27 July–1 August 1985	8	3	1
(same as last entry)	26 May 1986	3	2	–
1 mi. S + 1/2 mi. E Summerhaven, elev. 8000 ft., Santa Catalina Mts., Pima Co.	30 May 1987	–	2	–
(same as last entry)	3–4 April 1988	6	8	–
1 1/2 mi. W Hawk Peak, elev. 9700 ft., Pinaleno Mts., Graham Co.	25 May 1987	5	6	–
near Reef Mine, 9 mi. S Sierra Vista, elev. 7200 ft., Huachuca Mts., Cochise Co.	30 March 1988	2	3	–
Long Park, elev. 9200 ft., Chiricahua Mts., Cochise Co.	25 May 1985	2	–	–
Barfoot Park, elev. 8200 ft., Chiricahua Mts., Cochise Co.	22–28 May 1987	8	7	2
(same as last entry)	3–14 October 1988	8	7	–
Bootlegger Saddle, elev. 9000 ft., Chiricahua Mts., Cochise Co.	23 May 1989	1	2	3
Colorado and New Mexico				
White River Plateau, 10 mi. S + 2 mi. E Buford, elev. 10,000 ft., Rio Blanco Co., Colorado	5–7 August 1985	6	3	–
Tennessee Pass, elev. 10,400 ft., Eagle Co., Colorado	4–8 June 1986	16	9	1
Love Mesa, 16 mi. N + 7 mi. E Nucla, elev. 8100 ft., Mesa Co., Colorado	3 August 1985	1	–	–
Cottonwood Creek road, 17 mi. E Nucla, elev. 9200 ft., Montrose Co., Colorado	4 August 1985	–	1	–

Appendix A (continued)

Locality	Date	Males	Females	Juveniles[b]
Colorado and New Mexico (continued)				
Spring Creek Pass, elev. 11,100 ft, San Juan Mts., Hinsdale Co., Colorado	4–9 June 1987	8	10	1
McKenzie Ridge, 3 mi. NE McGaffey, elev. 8100 ft, Zuni Mts., McKinley Co., New Mexico	14–17 August 1987	13	6	1
Black Canyon, 24 mi. N + 3 mi. E Santa Rita, elev. 7500 ft, Black Mts., Grant Co., New Mexico	31 May 1987	2	1	–
Poverty Creek, 8 mi. N + 10 mi. W Chloride, elev. 7700 ft, Black Mts., Sierra Co., New Mexico	1 June 1987	5	3	–
Vick's Peak, 11 mi. N + 2 mi. E Monticello, elev. 8000 ft, San Mateo Mts., Socorro Co., New Mexico	2–3 June 1987	1	1	–
Southern Appalachians				
Brush Mountain, 2 mi. NW Blacksburg, elev. 2600 ft, Montgomery Co., Virginia	22 February–18 May 1983	25	13	–
(same as last entry)	19 March–20 May 1984	3	3	1
Poverty Hollow, 4 mi. NW Blacksburg, elev. 1900 ft, Montgomery Co., Virginia	27 May–17 December 1983	34	26	22
(same as last entry)	23 May–17 June 1984	9	3	3
Price Mountain, 2 mi. SE Blacksburg, elev. 2400 ft, Montgomery Co., Virginia	23 January–9 May 1984	1	1	1
Roan Mountain, elev. 6100 ft, Mitchell Co., North Carolina	20–21 June 1984	1	2	4
Highlands, elev. 3800 ft, Macon Co., North Carolina	30 October–21 November 1983	10	7	35
Northeastern Region				
6 mi. E Grand Marais, elev. 650 ft, Luce Co., Michigan	14–16 July 1987	8	3	8

a. Number of individuals of known morphology and flight calls.
b. Aged using plumage characters. Juveniles defined as members of plumage age classes 0–6 (see Materials and Methods).

Appendix B. Collections and Sample Sizes for Museum Specimens of *L. curvirostra* Examined in This Study

Collection	Acronym	Total	M[a]	F[b]
American Museum of Natural History[c]	AMNH	13	10	3
Burke Museum, University of Washington	BM	69	38	16
California Academy of Sciences	CAS	157	93	51
California State University, Long Beach	CSU	9	7	1
Carnegie Museum of Natural History	CMNH	272	111	64
Cornell University	CU	86	47	26
Delaware Museum of Natural History	DMNH	274	132	78
Denver Museum of Natural History	DM	125	67	37
Louisiana State University Museum of Zoology	LSUMZ	93	56	25
Museum of Vertebrate Zoology	MVZ	660	365	174
National Museum of Canada[c]	NMC	42	19	10
Royal British Columbia Museum[c]	RBCM	84	47	26
San Diego Natural History Museum	SDNHM	264	136	79
Santa Barbara Museum of Natural History	SBMNH	10	7	3
University of Arizona	UA	71	40	26
University of California, Los Angeles	UCLA	167	94	62
University of California, Santa Barbara	UCSB	1	—	1
University of Kansas Museum of Natural History	KU	242	133	90
University of Montana	UM	44	23	7
University of Nebraska State Museum	UNSM	69	40	28
University of Puget Sound	UPS	148	77	48
University of Wisconsin	UW	39	22	11
Washington State University	WSU	67	38	16

a. North American adult males.
b. North American adult females.
c. Only subsets of the total number of *L. curvirostra* study skin holdings were examined in these collections.

Appendix C. List of Recordings Made by Other Observers

Locality	Date	Call Types[a]	Observers[b]
Northwestern Region			
15 mi. W of Rapid City, South Dakota	10 June 1961	2	RSL, WYB
near Trapper Peak, Ravalli Co., Montana	3 July 1971	4	NKJ
Bismarck, North Dakota	May 1973	2	RNR
Seattle, Washington	4 February 1988	4	TH
Seattle, Washington	15–17 March 1988	3,4	TH
Fish Lake, Washington	15 November 1987	3,4	TH
Fish Lake, Washington	4 March 1988	2,4,5	TH
near Leavenworth, Washington	6 March 1988	2,3,4	TH
10 mi. N and 25 mi. W Yakima, Washington	29 July 1988	2,4	TH
near Neilton, Grays Harbor Co., Washington	29–30 August 1989	1,3	TH
Oregon			
near Burns, Oregon	12 May 1959	2	AAA, DGA
near Bend, Oregon	21 September 1984	2	EP
Sand Springs, near Bend, Oregon	9 December 1987	2	DJL
Northern and Coastal California			
Humboldt Co.	27 July 1984	4	MLM
Sierra Nevada of California			
North of Yosemite	2 June 1981	2	MLM
Yosemite National Park	27 July 1984	2	MLM
Arizona and southern Utah			
Devil's Canyon, Utah	26 June 1970	2	GBR
Pinaleno Mts., Arizona	17–18 July 1986	2,3,4	JTM
Colorado and New Mexico			
Jemez Falls, New Mexico	28 June 1983	2	CAS

Appendix C (continued)

Locality	Date	Call Types[a]	Observers[b]
Southern Appalachian Region			
Great Balsams, North Carolina	8 June 1987	1	DBM
Black Mts., North Carolina	10 and 23 June 1987	1,2	DBM
Great Smoky Mountains National Park	5 July 1981	1	CMN
Northeastern Region			
Acadia National Park, Maine	12 June 1962	4	RCS, RSL
Whitewater State Park, Indiana	February 1967	2	CSA
Rockford, Illinois	winter 1970	4	WMS
Cape Bonavista, Newfoundland	14 June 1981	8	JFP
Albany, New York	30 March 1982	2	CAS
Lake George, New York	6 May 1984	2	CWB
Acadia National Park, Maine	22–23 September 1984	4	CWB
Algonquin Park, Ontario	28–30 September 1984	2	CWB
Algonquin Park, Ontario	21 October 1984	2	CWB
Acadia National Park, Maine	30–31 October 1984	2,4	CWB
25 mi. S Albany, New York	4 February 1985	3	CWB
Hamilton Co., New York	14 February 1985	2,3,4	CWB
Alpena, Michigan	25 April 1985	3	RP
14 mi. E Grand Marais, Michigan	17 July 1986	2	RBP
near Timmins, Ontario	26 August 1987	3	CWB
Caledonia, Nova Scotia	12 October 1987	4	CWB

a. Numbers refer to call type designations given in the text.
b. Initials given for the following recordists: CSA = Curtis S. Adkisson, AAA = Arthur A. Allen, DGA = David G. Allen, CWB = Craig W. Benkman, WYB = W. Y. Brockleman, TH = Tom Hahn, NKJ = Ned K. Johnson, RSL = Randolph S. Little, DJL = Donna J. Lusthoff, MLM = Marie L. Mans, JTM = Joe T. Marshall, DBM = Douglas B. McNair, CN = Charles Nicholson, RBP = Robert B. Payne, JFP = Jay F. Pitocchelli, RP = Robert Preston, EP = Elinor Pugh, RNR = Robert N. Randall, GBR = George B. Reynard, WMS = William M. Shepherd, CAS = Cynthia A. Staicer, RCS = Robert C. Stein. Recordings by AAA, DGA, RCS, RSL, and WYB were provided by the Library of Natural Sounds, Laboratory of Ornithology, Cornell University, care of James L. Gulledge. Those by DBM and JFP were provided by the Florida State Museum, University of Florida, care of J. W. Hardy. The recording by NKJ is in the Collection of Natural Sounds, Museum of Vertebrate Zoology.

Appendix D. Univariate Measurements of Adults in 7 Vocal Groups of Red Crossbills

Body Mass

Group	Sex	N	Mean	SE	Range
Type 1	M	39	30.49	0.284	25.8–34.5
	F	33	29.09	0.389	24.5–35.1
Type 2	M	189	33.85	0.165	29.0–42.2
	F	118	32.88	0.237	26.3–41.4
Type 3	M	28	27.32	0.443	23.8–32.2
	F	14	26.54	0.469	23.7–29.4
Type 4	M	27	30.73	0.401	27.7–34.7
	F	18	28.56	0.708	24.2–35.0
Type 5	M	31	33.25	0.356	29.3–39.9
	F	24	32.85	0.410	29.5–38.1
Type 6	M	8	41.48	0.919	38.5–45.4
	F	7	39.43	0.913	36.9–42.4
Type 7	M	5	31.56	1.190	29.1–36.0
	F	1	26.9	—	—

Toe Length

Group	Sex	N	Mean	SE	Range
Type 1	M	12	10.89	0.128	10.1–11.5
	F	5	10.74	0.136	10.3–11.1
Type 2	M	152	11.46	0.037	10.3–13.3
	F	95	11.11	0.053	10.1–12.3
Type 3	M	27	10.92	0.066	10.3–11.6
	F	16	10.78	0.071	10.3–11.3
Type 4	M	27	11.13	0.083	10.4–12.2
	F	18	10.85	0.123	9.7–11.6
Type 5	M	29	11.40	0.062	10.8–12.1
	F	24	10.93	0.084	10.1–11.8
Type 6	M	8	12.41	0.120	11.9–12.9
	F	6	12.45	0.145	12.1–13.1
Type 7	M	5	11.36	0.194	10.9–11.9
	F	1	11.2	—	—

Tail Length

Group	Sex	N	Mean	SE	Range
Type 1	M	12	53.86	0.773	49.9–58.4
	F	5	50.92	0.974	48.2–53.6
Type 2	M	152	56.03	0.159	52.0–61.3
	F	95	54.39	0.181	50.9–57.9
Type 3	M	27	51.55	0.303	48.2–55.4
	F	16	49.31	0.384	46.6–52.1
Type 4	M	27	53.79	0.332	48.8–56.7
	F	18	51.81	0.374	49.2–55.5
Type 5	M	29	57.93	0.307	55.4–63.2
	F	24	55.93	0.290	52.6–58.8
Type 6	M	8	57.94	0.400	56.5–59.5
	F	6	57.63	0.664	55.6–59.8
Type 7	M	5	55.48	0.830	53.6–57.7
	F	1	52.3	—	—

Tarsus Length

Group	Sex	N	Mean	SE	Range
Type 1	M	39	19.16	0.090	18.0–20.4
	F	33	18.90	0.103	17.3–20.3
Type 2	M	189	20.19	0.044	18.6–21.8
	F	118	19.95	0.064	17.5–21.4
Type 3	M	28	18.80	0.127	17.2–19.9
	F	17	18.47	0.109	17.5–19.4
Type 4	M	28	19.44	0.119	18.1–21.0
	F	19	19.02	0.095	18.4–19.9
Type 5	M	31	20.14	0.097	19.1–21.5
	F	24	20.14	0.122	18.7–21.6
Type 6	M	8	21.35	0.258	20.4–22.4
	F	7	21.11	0.182	20.5–21.6
Type 7	M	5	20.48	0.188	19.9–20.9
	F	1	19.2	—	—

Wing Length

Group	Sex	N	Mean	SE	Range
Type 1	M	39	89.41	0.374	83.8–95.4
	F	33	86.50	0.395	82.1–90.3
Type 2	M	189	93.78	0.145	89.2–98.7
	F	117	90.56	0.186	85.8–96.0
Type 3	M	28	85.88	0.343	81.8–88.5
	F	17	82.37	0.518	78.0–85.5
Type 4	M	28	90.90	0.384	85.0–94.8
	F	19	86.78	0.454	83.6–90.0
Type 5	M	31	95.38	0.244	92.4–98.0
	F	24	91.99	0.336	88.8–95.0
Type 6	M	8	98.36	0.570	95.5–101.2
	F	7	97.20	0.348	95.7–98.1
Type 7	M	5	93.18	0.357	92.0–94.1
	F	1	90.0	—	—

Upper Mandible Length

Group	Sex	N	Mean	SE	Range
Type 1	M	39	14.49	0.101	13.3–15.7
	F	33	14.19	0.099	12.9–15.0
Type 2	M	188	16.14	0.059	14.6–19.8
	F	118	15.68	0.076	13.2–18.1
Type 3	M	28	12.74	0.117	11.7–14.3
	F	17	12.81	0.126	11.8–13.7
Type 4	M	28	14.60	0.105	13.4–16.0
	F	19	14.34	0.200	13.4–17.2
Type 5	M	31	15.53	0.135	14.6–17.7
	F	24	15.30	0.144	13.8–16.8
Type 6	M	8	17.14	0.209	16.2–17.9
	F	7	16.84	0.318	15.6–18.2
Type 7	M	5	14.76	0.317	13.6–15.5
	F	1	15.2	—	—

Appendix D (continued)

Lower Mandible Length

Group	Sex	N	Mean	SE	Range
Type 1	M	39	11.32	0.088	10.5–12.9
	F	33	11.09	0.082	10.3–12.4
Type 2	M	189	12.67	0.053	11.3–15.5
	F	118	12.37	0.061	10.9–14.3
Type 3	M	28	10.38	0.091	9.4–11.5
	F	17	9.94	0.109	9.2–10.6
Type 4	M	28	11.48	0.101	10.5–12.6
	F	19	11.31	0.173	10.3–13.5
Type 5	M	31	12.25	0.121	11.4–14.6
	F	24	11.85	0.074	10.8–12.6
Type 6	M	8	14.09	0.202	13.0–14.8
	F	7	13.69	0.190	13.0–14.6
Type 7	M	5	11.80	0.176	11.3–12.2
	F	1	11.2	—	—

Upper Mandible Depth

Group	Sex	N	Mean	SE	Range
Type 1	M	39	5.08	0.021	4.8–5.4
	F	33	5.05	0.020	4.8–5.3
Type 2	M	189	5.56	0.012	5.2–6.1
	F	118	5.43	0.018	5.0–6.0
Type 3	M	28	4.74	0.025	4.5–4.9
	F	17	4.72	0.031	4.5–4.9
Type 4	M	28	5.21	0.026	4.8–5.5
	F	19	4.98	0.032	4.7–5.3
Type 5	M	31	5.55	0.033	5.3–6.0
	F	24	5.44	0.033	5.1–5.7
Type 6	M	8	6.23	0.059	6.0–6.5
	F	7	6.07	0.068	5.8–6.3
Type 7	M	5	5.54	0.040	5.4–5.6
	F	1	5.4	—	—

Upper Mandible Width

Group	Sex	N	Mean	SE	Range
Type 1	M	39	6.46	0.036	6.0–7.0
	F	33	6.44	0.039	6.0–6.9
Type 2	M	189	7.28	0.018	6.7–7.9
	F	118	7.18	0.027	6.4–7.9
Type 3	M	28	6.09	0.041	5.6–6.5
	F	17	5.95	0.068	5.4–6.4
Type 4	M	28	6.63	0.062	6.0–7.3
	F	19	6.46	0.055	6.1–7.0
Type 5	M	31	7.34	0.046	6.6–7.8
	F	24	7.20	0.046	6.8–7.5
Type 6	M	8	8.21	0.058	8.0–8.4
	F	7	7.90	0.076	7.7–8.2
Type 7	M	5	7.28	0.139	6.8–7.6
	F	1	6.8	—	—

Bill Depth

Group	Sex	N	Mean	SE	Range
Type 1	M	39	8.80	0.041	8.3–9.4
	F	33	8.72	0.040	8.3–9.2
Type 2	M	189	9.67	0.026	8.9–10.6
	F	118	9.41	0.029	8.7–10.5
Type 3	M	28	8.19	0.046	7.8–8.7
	F	17	8.10	0.055	7.7–8.5
Type 4	M	28	9.00	0.040	8.4–9.4
	F	19	8.52	0.057	7.8–9.0
Type 5	M	31	9.57	0.056	8.9–10.2
	F	24	9.28	0.047	8.9–9.7
Type 6	M	8	11.11	0.117	10.7–11.7
	F	7	10.86	0.134	10.2–11.3
Type 7	M	5	9.56	0.108	9.3–9.9
	F	1	9.5	—	—

Diagonal Upper Mandible Depth

Group	Sex	N	Mean	SE	Range
Type 1	M	4	5.67	0.149	5.4–6.1
	F	3	5.57	0.120	5.4–5.8
Type 2	M	149	6.30	0.018	5.8–7.0
	F	94	6.17	0.025	5.7–6.8
Type 3	M	27	5.46	0.043	5.1–5.9
	F	16	5.44	0.058	5.0–5.8
Type 4	M	27	5.94	0.031	5.6–6.2
	F	18	5.74	0.058	5.3–6.2
Type 5	M	29	6.27	0.042	5.9–6.8
	F	24	6.10	0.033	5.7–6.4
Type 6	M	8	6.91	0.091	6.4–7.3
	F	6	6.70	0.052	6.5–6.8
Type 7	M	5	6.14	0.150	5.8–6.7
	F	1	6.1	—	—

Lower Mandible Width

Group	Sex	N	Mean	SE	Range
Type 1	M	39	9.81	0.048	9.1–10.4
	F	33	9.80	0.064	8.9–10.5
Type 2	M	189	10.54	0.026	9.4–11.6
	F	118	10.26	0.040	9.3–11.5
Type 3	M	28	8.86	0.067	8.1–9.8
	F	17	8.69	0.062	8.4–9.2
Type 4	M	28	9.78	0.075	9.1–10.7
	F	19	9.36	0.064	8.9–9.9
Type 5	M	31	10.29	0.058	9.6–11.0
	F	24	10.02	0.064	9.3–10.5
Type 6	M	8	11.81	0.111	11.4–12.2
	F	7	11.40	0.151	10.9–11.9
Type 7	M	5	10.34	0.181	9.7–10.8
	F	1	9.7	—	—

Appendix D (continued)

Curvature Angle (in degrees)

Group	Sex	N	Mean	SE	Range
Type 1	M	4	18.87	0.483	17.8–19.7
	F	3	17.99	0.343	17.3–18.3
Type 2	M	147	18.36	0.147	13.2–22.0
	F	94	18.37	0.190	13.1–22.0
Type 3	M	28	18.09	0.340	15.1–22.3
	F	16	18.57	0.410	15.8–21.2
Type 4	M	27	18.07	0.380	13.9–23.2
	F	18	18.17	0.376	15.8–21.0
Type 5	M	29	17.63	0.281	12.7–20.7
	F	24	17.53	0.290	14.6–20.1
Type 6	M	8	20.45	0.540	18.5–23.2
	F	6	20.92	0.509	18.9–22.1
Type 7	M	5	17.40	0.723	14.9–19.1
	F	1	16.3	—	—

Skull Width

Group	Sex	N	Mean	SE	Range
Type 1	M	13	17.85	0.137	16.9–18.5
	F	10	17.27	0.165	16.4–18.1
Type 2	M	153	18.37	0.031	17.6–19.5
	F	94	17.88	0.074	12.3–19.0
Type 3	M	28	16.73	0.060	16.1–17.3
	F	18	16.52	0.063	16.0–16.9
Type 4	M	28	17.55	0.064	17.0–18.2
	F	17	17.01	0.105	16.4–17.9
Type 5	M	31	18.38	0.069	17.6–19.0
	F	24	18.06	0.086	17.0–18.8
Type 6	M	8	19.64	0.143	19.1–20.2
	F	7	19.19	0.158	18.6–19.8
Type 7	M	5	18.22	0.066	18.0–18.4
	F	1	18.0	—	—

Sclerotic Ring Width

Group	Sex	N	Mean	SE	Range
Type 1	M	0	—	—	—
	F	1	6.3	—	—
Type 2	M	123	6.44	0.015	6.0–6.8
	F	70	6.41	0.022	6.0–6.7
Type 3	M	22	6.07	0.032	5.8–6.4
	F	16	6.07	0.044	5.8–6.5
Type 4	M	23	6.31	0.036	6.0–6.6
	F	14	6.14	0.048	5.8–6.4
Type 5	M	28	6.44	0.024	6.1–6.7
	F	24	6.37	0.028	6.1–6.6
Type 6	M	8	6.82	0.049	6.6–7.0
	F	7	6.79	0.059	6.5–7.0
Type 7	M	4	6.30	0.041	6.2–6.4
	F	1	6.6	—	—

Postorbital Width

Group	Sex	N	Mean	SE	Range
Type 1	M	13	15.52	0.106	14.9–16.2
	F	10	15.19	0.105	14.7–15.7
Type 2	M	154	15.84	0.024	15.0–16.7
	F	93	15.62	0.031	14.8–16.6
Type 3	M	28	15.08	0.047	14.5–15.6
	F	18	14.96	0.076	14.2–15.5
Type 4	M	28	15.53	0.069	14.6–16.1
	F	17	15.12	0.081	14.5–15.8
Type 5	M	31	16.00	0.064	15.4–16.9
	F	24	15.82	0.070	15.2–16.5
Type 6	M	8	16.48	0.121	15.9–16.9
	F	7	16.40	0.179	15.8–16.9
Type 7	M	5	15.84	0.136	15.3–16.0
	F	1	15.6	—	—

Skull Depth

Group	Sex	N	Mean	SE	Range
Type 1	M	13	12.91	0.072	12.5–13.3
	F	10	12.84	0.105	12.3–13.4
Type 2	M	153	13.36	0.019	12.6–14.1
	F	94	13.14	0.028	12.5–13.9
Type 3	M	27	12.59	0.048	12.0–13.3
	F	18	12.44	0.052	12.1–12.9
Type 4	M	28	13.00	0.058	12.5–13.6
	F	17	12.79	0.065	12.2–13.3
Type 5	M	31	13.37	0.048	12.8–13.9
	F	24	13.19	0.051	12.8–13.7
Type 6	M	8	14.15	0.113	13.6–14.5
	F	7	13.69	0.108	13.2–14.0
Type 7	M	5	13.30	0.071	13.1–13.5
	F	1	13.0	—	—

Coracoid Width

Group	Sex	N	Mean	SE	Range
Type 1	M	13	1.087	0.019	1.00–1.26
	F	10	1.051	0.022	0.96–1.15
Type 2	M	155	1.098	0.005	0.94–1.34
	F	94	1.075	0.007	0.95–1.25
Type 3	M	28	1.029	0.008	0.95–1.12
	F	18	1.013	0.012	0.93–1.08
Type 4	M	28	1.101	0.012	0.98–1.24
	F	18	1.038	0.013	0.91–1.11
Type 5	M	31	1.069	0.011	0.95–1.16
	F	24	1.070	0.011	0.96–1.17
Type 6	M	8	1.091	0.023	1.00–1.21
	F	7	1.113	0.021	1.04–1.21
Type 7	M	5	1.054	0.024	1.00–1.14
	F	1	1.10	—	—

Appendix D (continued)

Coracoid Length

Group	Sex	N	Mean	SE	Range
Type 1	M	13	20.08	0.118	19.5–20.9
	F	10	19.66	0.118	19.2–20.2
Type 2	M	155	20.92	0.043	19.2–22.2
	F	94	20.33	0.062	18.7–21.7
Type 3	M	28	19.30	0.061	18.6–19.8
	F	17	18.92	0.128	17.8–19.8
Type 4	M	28	20.08	0.110	19.0–21.6
	F	18	19.32	0.133	18.1–20.3
Type 5	M	31	20.71	0.089	19.7–21.8
	F	24	20.22	0.080	19.3–20.9
Type 6	M	8	21.98	0.217	21.2–22.9
	F	7	22.03	0.166	21.2–22.6
Type 7	M	5	20.40	0.257	19.5–20.9
	F	1	18.90	—	—

Scapula Length

Group	Sex	N	Mean	SE	Range
Type 1	M	39	21.95	0.187	21.0–23.0
	F	9	21.69	0.238	20.7–22.9
Type 2	M	153	22.94	0.055	21.1–24.9
	F	94	22.37	0.077	20.2–24.2
Type 3	M	28	20.99	0.094	20.0–21.8
	F	18	20.56	0.137	19.8–21.6
Type 4	M	28	22.08	0.125	20.9–23.5
	F	18	21.01	0.187	19.5–22.6
Type 5	M	31	22.57	0.105	21.6–23.9
	F	24	22.31	0.136	20.8–23.9
Type 6	M	8	24.31	0.255	23.6–25.9
	F	7	24.09	0.251	23.1–25.0
Type 7	M	5	22.58	0.273	21.5–23.0
	F	1	22.2	—	—

Humerus Length

Group	Sex	N	Mean	SE	Range
Type 1	M	13	18.54	0.126	17.9–19.5
	F	9	18.11	0.087	17.6–18.4
Type 2	M	155	19.24	0.035	18.2–20.3
	F	95	18.84	0.052	17.7–20.0
Type 3	M	28	17.80	0.061	17.1–18.3
	F	18	17.37	0.107	16.6–18.6
Type 4	M	28	18.47	0.078	17.4–19.3
	F	18	17.92	0.092	17.3–18.5
Type 5	M	30	19.17	0.073	18.4–20.3
	F	24	18.72	0.079	17.9–19.3
Type 6	M	8	20.43	0.144	19.7–20.9
	F	7	20.36	0.129	19.9–20.8
Type 7	M	5	19.04	0.144	18.6–19.3
	F	1	18.9	—	—

Scapula Width

Group	Sex	N	Mean	SE	Range
Type 1	M	13	4.49	0.045	4.2–4.7
	F	10	4.40	0.049	4.2–4.6
Type 2	M	156	4.69	0.012	4.3–5.2
	F	95	4.57	0.018	4.1–5.0
Type 3	M	28	4.30	0.024	4.1–4.7
	F	18	4.19	0.022	4.0–4.3
Type 4	M	28	4.53	0.024	4.2–4.7
	F	18	4.29	0.051	3.9–4.7
Type 5	M	31	4.62	0.029	4.2–4.9
	F	24	4.45	0.039	4.1–4.8
Type 6	M	8	5.00	0.068	4.7–5.2
	F	7	4.83	0.036	4.7–5.0
Type 7	M	5	4.58	0.049	4.5–4.7
	F	1	4.6	—	—

Humerus Width

Group	Sex	N	Mean	SE	Range
Type 1	M	13	1.632	0.019	1.51–1.78
	F	10	1.580	0.011	1.53–1.63
Type 2	M	155	1.679	0.005	1.50–1.88
	F	95	1.652	0.007	1.49–1.87
Type 3	M	28	1.565	0.012	1.46–1.70
	F	18	1.544	0.012	1.46–1.63
Type 4	M	28	1.648	0.012	1.52–1.78
	F	18	1.587	0.020	1.45–1.79
Type 5	M	30	1.653	0.010	1.55–1.76
	F	24	1.636	0.010	1.56–1.72
Type 6	M	8	1.787	0.025	1.71–1.94
	F	7	1.751	0.022	1.65–1.82
Type 7	M	5	1.634	0.016	1.60–1.69
	F	1	1.74	—	—

Sternum Length

Group	Sex	N	Mean	SE	Range
Type 1	M	13	23.30	0.147	22.1–24.0
	F	10	23.05	0.256	20.8–23.6
Type 2	M	155	24.33	0.057	22.3–26.8
	F	94	23.56	0.078	21.5–25.9
Type 3	M	28	22.31	0.109	21.2–23.3
	F	18	21.67	0.162	20.3–22.8
Type 4	M	28	23.44	0.127	22.2–24.7
	F	18	22.24	0.136	21.5–23.3
Type 5	M	30	24.05	0.127	22.6–25.6
	F	24	23.54	0.093	22.6–24.4
Type 6	M	8	25.90	0.252	24.6–26.7
	F	7	25.03	0.225	23.9–25.7
Type 7	M	5	23.54	0.211	23.8–24.8
	F	1	23.0	—	—

Appendix D (continued)

Keel Length

Group	Sex	N	Mean	SE	Range
Type 1	M	13	23.57	0.250	21.6–25.2
	F	10	22.77	0.296	20.5–23.9
Type 2	M	152	24.99	0.069	22.6–27.6
	F	94	23.55	0.078	21.0–25.9
Type 3	M	28	23.00	0.165	20.8–24.7
	F	17	21.84	0.201	20.5–23.0
Type 4	M	28	24.20	0.137	22.7–25.7
	F	18	22.17	0.183	20.9–23.8
Type 5	M	31	24.67	0.139	23.0–26.2
	F	24	23.50	0.136	22.1–24.7
Type 6	M	8	26.49	0.288	25.4–28.0
	F	7	25.21	0.275	24.3–26.3
Type 7	M	5	24.28	0.185	23.8–24.8
	F	1	22.8	—	—

Keel Depth

Group	Sex	N	Mean	SE	Range
Type 1	M	13	11.99	0.102	11.5–12.8
	F	10	11.91	0.117	11.4–12.6
Type 2	M	152	12.33	0.036	11.2–13.5
	F	93	12.05	0.045	10.8–13.4
Type 3	M	28	11.28	0.076	10.4–12.2
	F	17	11.35	0.100	10.7–11.8
Type 4	M	28	12.02	0.062	11.3–12.6
	F	18	11.61	0.067	11.1–12.1
Type 5	M	30	12.07	0.070	11.3–12.6
	F	24	11.88	0.065	11.2–12.6
Type 6	M	8	12.66	0.158	12.2–13.4
	F	7	12.33	0.144	11.7–12.7
Type 7	M	5	12.06	0.103	11.7–12.3
	F	1	11.6	—	—

Anterior Synsacrum Length

Group	Sex	N	Mean	SE	Range
Type 1	M	13	11.25	0.076	10.7–11.6
	F	10	10.92	0.112	10.2–11.5
Type 2	M	152	11.44	0.026	10.8–12.5
	F	94	11.22	0.041	10.4–12.2
Type 3	M	28	10.55	0.074	9.7–11.6
	F	18	10.56	0.080	10.1–11.4
Type 4	M	28	11.04	0.065	10.5–11.8
	F	18	10.78	0.078	10.3–11.4
Type 5	M	31	11.41	0.069	10.7–12.2
	F	24	11.30	0.071	10.6–11.9
Type 6	M	8	12.28	0.179	11.5–12.9
	F	7	12.23	0.123	11.9–12.7
Type 7	M	5	11.04	0.129	10.6–11.3
	F	1	—	—	—

Sternum Width

Group	Sex	N	Mean	SE	Range
Type 1	M	13	12.07	0.079	11.7–12.6
	F	10	12.13	0.107	11.5–12.6
Type 2	M	151	12.57	0.040	10.8–14.3
	F	92	12.43	0.048	11.4–14.0
Type 3	M	27	11.61	0.067	11.0–12.3
	F	16	11.49	0.104	10.7–12.2
Type 4	M	28	12.15	0.079	11.5–13.3
	F	17	11.85	0.083	11.2–12.4
Type 5	M	29	12.41	0.082	11.7–13.5
	F	23	12.13	0.086	11.2–13.1
Type 6	M	7	13.16	0.075	12.9–13.5
	F	6	12.85	0.232	12.2–13.8
Type 7	M	5	12.22	0.258	11.5–12.9
	F	1	12.4	—	—

Furcular Process Length

Group	Sex	N	Mean	SE	Range
Type 1	M	13	5.88	0.107	5.3–6.4
	F	10	5.47	0.143	4.7–6.1
Type 2	M	153	5.89	0.033	5.0–6.8
	F	91	5.49	0.049	4.5–6.9
Type 3	M	27	5.42	0.075	4.6–6.1
	F	18	5.23	0.105	4.6–6.0
Type 4	M	28	5.88	0.094	4.9–6.7
	F	17	5.21	0.106	4.3–6.1
Type 5	M	30	5.74	0.075	5.1–6.5
	F	24	5.43	0.104	4.5–6.2
Type 6	M	8	5.89	0.138	5.4–6.6
	F	7	5.73	0.157	5.2–6.5
Type 7	M	5	5.98	0.242	5.4–6.6
	F	1	5.0	—	—

Synsacrum Minimum Width

Group	Sex	N	Mean	SE	Range
Type 1	M	13	7.40	0.105	6.9–8.0
	F	10	7.39	0.097	6.8–7.8
Type 2	M	147	7.68	0.030	6.8–8.7
	F	91	7.68	0.036	7.0–8.6
Type 3	M	28	7.19	0.059	6.7–8.0
	F	18	7.24	0.050	6.9–7.6
Type 4	M	28	7.42	0.072	6.9–8.4
	F	18	7.30	0.070	6.8–7.8
Type 5	M	31	7.69	0.067	7.0–8.4
	F	24	7.66	0.067	7.0–8.4
Type 6	M	8	8.20	0.120	7.5–8.6
	F	7	7.97	0.141	7.4–8.6
Type 7	M	5	7.68	0.080	7.5–7.9
	F	1	7.9	—	—

Appendix D (continued)

Synsacrum Width

Group	Sex	N	Mean	SE	Range
Type 1	M	13	12.35	0.105	11.7–12.9
	F	9	12.46	0.174	11.6–13.0
Type 2	M	147	12.84	0.036	11.3–13.9
	F	91	12.70	0.043	11.7–13.7
Type 3	M	27	11.88	0.073	11.3–12.8
	F	17	11.91	0.076	11.4–12.4
Type 4	M	28	12.46	0.088	11.2–13.4
	F	18	12.07	0.083	11.4–12.7
Type 5	M	30	12.72	0.073	11.9–13.7
	F	22	12.72	0.090	11.6–13.7
Type 6	M	8	13.44	0.159	12.6–14.1
	F	7	13.59	0.134	13.0–14.1
Type 7	M	5	12.74	0.175	12.2–13.1
	F	1	13.1	—	—

Femur Length

Group	Sex	N	Mean	SE	Range
Type 1	M	13	18.02	0.118	17.3–18.6
	F	10	17.85	0.130	17.1–18.5
Type 2	M	151	18.68	0.033	17.7–19.8
	F	95	18.55	0.055	16.9–19.8
Type 3	M	28	17.23	0.071	16.3–17.8
	F	18	17.14	0.106	16.4–18.1
Type 4	M	28	17.98	0.094	16.8–19.0
	F	18	17.67	0.108	16.8–18.3
Type 5	M	30	18.50	0.071	17.7–19.3
	F	24	18.38	0.087	17.6–19.1
Type 6	M	7	19.76	0.177	19.0–20.4
	F	7	19.71	0.193	19.0–20.4
Type 7	M	5	18.70	0.134	18.4–19.2
	F	1	19.1	—	—

Tibiotarsus Length

Group	Sex	N	Mean	SE	Range
Type 1	M	13	27.86	0.202	26.4–29.1
	F	10	27.83	0.221	26.8–28.8
Type 2	M	148	29.03	0.060	27.2–30.8
	F	90	28.82	0.092	26.1–31.4
Type 3	M	27	27.11	0.111	25.7–27.9
	F	18	26.96	0.143	25.9–28.2
Type 4	M	28	27.94	0.149	25.7–29.3
	F	18	27.54	0.178	25.6–28.8
Type 5	M	31	28.75	0.142	27.1–30.7
	F	23	28.84	0.102	27.5–29.8
Type 6	M	8	30.83	0.247	29.5–31.7
	F	7	31.06	0.230	30.0–31.9
Type 7	M	5	28.86	0.293	28.0–29.6
	F	1	29.3	—	—

Femur Width

Group	Sex	N	Mean	SE	Range
Type 1	M	13	1.376	0.019	1.26–1.49
	F	10	1.316	0.009	1.27–1.35
Type 2	M	152	1.395	0.004	1.27–1.51
	F	95	1.374	0.006	1.22–1.52
Type 3	M	28	1.305	0.012	1.17–1.42
	F	18	1.295	0.006	1.25–1.35
Type 4	M	28	1.364	0.010	1.22–1.46
	F	18	1.301	0.014	1.20–1.40
Type 5	M	30	1.395	0.010	1.29–1.50
	F	24	1.377	0.008	1.31–1.48
Type 6	M	7	1.494	0.013	1.45–1.54
	F	7	1.476	0.015	1.40–1.53
Type 7	M	5	1.364	0.011	1.34–1.39
	F	1	1.46	—	—

Tibiotarsus Width

Group	Sex	N	Mean	SE	Range
Type 1	M	13	1.212	0.022	1.12–1.39
	F	10	1.220	0.021	1.08–1.31
Type 2	M	148	1.271	0.005	1.10–1.48
	F	91	1.240	0.006	1.10–1.37
Type 3	M	27	1.180	0.013	1.02–1.30
	F	18	1.189	0.009	1.13–1.27
Type 4	M	28	1.240	0.011	1.09–1.33
	F	18	1.191	0.011	1.08–1.28
Type 5	M	31	1.249	0.013	1.11–1.41
	F	23	1.246	0.010	1.16–1.33
Type 6	M	8	1.340	0.019	1.23–1.41
	F	7	1.351	0.027	1.26–1.45
Type 7	M	5	1.224	0.021	1.18–1.30
	F	1	1.30	—	—

Appendix E. Individual Identification Numbers[a], Catalog Numbers[b], and Vocal Group Numbers[c] for the Sample of Red Crossbills with Recorded Vocalizations

Identification Number	Catalog Number	Vocal Group	Identification Number	Catalog Number	Vocal Group	Identification Number	Catalog Number	Vocal Group
aF1	60	1	aF53	nc	1	jM111	nc	2
aF2	nc	1	aM55	nc	2	aF112	nc	2
aF3	nc	1	aM56	nc	1	aM113	66	2
aF4	99	1	aM57	nc	1	aF114	nc	1
aM5	nc	1	aM58	nc	1	jM115	nc	2
aF6	nc	1	aF59	nc	1	jF116	nc	2
aM7	410	2	aF60	nc	2	jF117	nc	2
aM8	nc	1	aF61	nc	1	jM118	nc	2
aF9	nc	1	aM62	nc	2	jM119	nc	2
aM10	238	2	aF63	nc	1	jM120	15	2
aM11	71	1	aM67	nc	2	jM121	67	1
aM12	73	2	aF69	nc	2	jF122	68	1
aM13	nc	1	aF70	nc	2	jM123	nc	1
aF14	nc	2	aM71	nc	2	jF124	nc	2
aM15	61	1	aM72	nc	2	aF125	35	2
aM16	nc	2	aF73	625	2	aF126	18	2
aF17	nc	1	aM74	nc	1	jM127	19	1
aM18	nc	1	aF75	nc	1	jM128	nc	1
aM19	74	1	jM76	5	2	aM129	20	1
aM20	nc	2	aM77	nc	2	jF130	21	1
aM21	nc	2	aM78	nc	2	jM131	nc	1
aM22	nc	2	aM79	nc	2	jM132	22	1
aM23	nc	2	jF80	nc	2	jM133	23	1
aF24	208	1	aM81	nc	2	aF134	24	1
aF25	62	1	aF82	nc	2	aM135	25	1
aM26	718	1	aM83	nc	2	jF136	26	1
aM27	nc	1	aF84	nc	2	jF137	nc	1
aM28	nc	2	aM85	nc	1	jM138	27	2
aM29	nc	1	aM86	nc	2	jM139	28	1
aF30	nc	1	aF87	nc	2	jF140	nc	1
aM31	nc	1	aM88	9	2	jM141	30	1
aM32	nc	1	aF89	nc	1	jM142	nc	1
aM33	nc	1	aM90	nc	1	aM143	31	1
aF34	nc	1	aM91	nc	2	jF144	nc	1
aM35	nc	1	aM92	nc	2	jM145	nc	1
aM36	nc	2	aF93	nc	2	jF146	32	1
aM37	63	2	jM94	nc	2	aF147	33	1
aF38	75	2	jM95	nc	2	jM149	nc	1
jM39	1	2	jM97	nc	2	jM150	nc	1
aM40	nc	1	aF98	11	2	jM151	nc	1
aM42	nc	2	aM99	12	2	jF152	nc	1
aM43	70	2	aF100	13	2	jF153	nc	1
aF44	nc	2	aM101	14	2	jM154	34	1
aF46	nc	1	jM103	nc	2	jM155	nc	1
aF47	nc	2	aM104	nc	2	jM156	nc	1
aM48	nc	2	aF105	nc	2	jF157	nc	1
aM49	nc	2	aF106	nc	2	aM158	nc	1
aF50	64	1	aM107	nc	2	jF159	nc	1
aF51	nc	1	aM108	nc	2	aM160	nc	1
aM52	nc	1	jM110	nc	2	aM161	nc	1

Appendix E (continued)

Identification Number	Catalog Number	Vocal Group	Identification Number	Catalog Number	Vocal Group	Identification Number	Catalog Number	Vocal Group
jM162	nc	1	jM213	91	1	aF263	131	2
jM163	nc	1	aF214	92	1	aM264	141	2
aF164	nc	1	aF215	93	1	aM265	130	2
jM165	nc	1	aM216	94	1	aF266	129	2
aM166	nc	1	aM217	109	3	aF267	148	3
jM167	nc	1	aM218	100	3	aF268	149	2
jF168	nc	1	aM219	101	3	aM269	150	3
aF169	nc	1	aM220	102	3	aF270	151	3
aF170	nc	1	jM221	103	3	aF271	152	3
aM171	nc	1	jO222	104	3	aM272	153	3
aF172	nc	1	aF223	105	3	aM273	600	3
aM173	nc	1	aF224	106	3	aF274	626	3
jM174	nc	1	aF225	107	3	aM275	154	2
aM175	nc	1	aM226	108	3	aM276	155	2
jF176	nc	1	aF227	110	3	aM277	156	2
aF177	nc	1	aF228	157	2	aM278	159	2
jM178	nc	1	aF229	158	2	aF279	160	2
jM179	nc	1	aF230	116	2	aF280	162	3
aM180	nc	2	aF231	118	2	aM281	165	3
aF181	nc	2	aF232	128	2	aF282	166	3
jF182	nc	2	aM233	123	2	aM283	164	3
jM183	nc	2	aF234	119	2	aM284	163	3
jM184	36	2	aM235	111	2	aM285	167	3
aM185	37	1	aF236	121	2	aM286	171	4
aF186	38	1	aM237	122	2	aM287	172	4
aF188	69	1	aM238	127	2	aM290	175	2
aM189	nc	2	aM239	124	2	aM291	176	2
aF190	nc	2	aF240	114	2	aM292	179	2
aF191	nc	1	aF241	126	2	aM293	180	2
aM192	40	1	aF242	145	2	aM294	174	3
jM193	nc	2	aM243	146	2	aF295	177	3
aM194	41	1	aF244	125	2	aM296	178	3
aM195	42	1	aF245	147	2	aM297	181	3
aF196	47	2	aM246	134	2	aF298	694	3
aM197	48	1	aM247	117	2	aM299	182	3
aM198	50	1	aM248	115	2	aM303	184	4
aM199	51	1	aM249	113	2	aM304	185	4
aF200	52	1	aM250	135	2	aM305	567	4
aM201	53	2	aM251	136	2	aF306	722	4
aF202	57	1	aM252	120	2	aF307	186	4
aM203	58	2	aM253	112	2	aF308	187	4
aM204	59	1	aM254	137	2	aM309	188	4
aM205	76	2	aF255	133	2	aM310	189	4
aM206	77	2	aM256	144	2	aM311	190	3
jM207	82	2	aM257	138	2	aM312	191	4
jM208	83	2	aF258	143	3	aF313	192	4
jM209	84	2	aM259	142	3	aM316	212	2
jF210	88	2	aM260	132	2	aM317	211	4
jF211	89	1	aF261	139	2	aM319	213	3
jF212	90	1	aM262	140	2	aM320	214	4

Appendix E (continued)

Identification Number	Catalog Number	Vocal Group	Identification Number	Catalog Number	Vocal Group	Identification Number	Catalog Number	Vocal Group
aF321	217	4	aM373	272	2	aF428	328	4
aF322	215	4	aF374	nc	2	aM429	329	4
aF323	220	3	aM375	279	2	jM430	325	4
aM324	218	3	aF376	283	2	jF431	326	4
aM325	219	4	aM377	282	2	jF432	336	4
aM326	216	4	aF378	280	2	jF433	338	3
aM327	273	2	aF379	277	2	jF434	333	4
aM328	224	2	aF380	278	2	jF435	331	4
aF330	221	2	aF381	276	2	aM436	337	3
aM331	222	2	jF382	281	2	aM437	339	4
aM332	226	2	jM383	285	2	jM438	335	3
aM333	223	2	jF384	691	2	aM439	334	4
aF334	231	2	jM385	709	2	aM440	330	4
aM335	227	2	jM386	284	2	aM441	327	4
aF336	228	2	aF387	568	2	aF442	332	4
aM337	229	2	aF388	286	2	jF443	341	4
aM338	230	2	aF389	288	2	jM444	342	4
aF339	232	4	aM390	292	2	jF445	348	4
aF340	233	3	aM392	291	2	jM446	340	4
aM341	235	4	aM393	290	2	jF447	344	4
aF342	234	4	aF394	289	2	aF448	345	3
aF343	236	2	aM395	693	5	aM449	346	3
aM344	237	2	aM396	295	5	aM450	343	4
aM347	239	2	aF397	293	5	jF451	347	3
aM348	247	2	aF398	294	5	jF452	349	2
aM349	240	2	aM399	296	5	jF453	350	2
aF350	243	2	aM400	297	2	jF454	356	2
aM351	244	2	aM401	298	5	jF455	355	2
aM352	241	2	aF402	299	5	jF456	353	4
jM353	242	2	aF403	300	5	jM457	352	4
aF354	245	2	aM404	303	2	jF458	354	2
aM355	246	2	aM405	301	2	jM459	357	2
aF356	251	2	aM406	302	5	jF460	351	4
aM357	248	2	aF407	304	5	aM461	359	4
aM358	250	2	aF408	305	5	aM462	358	2
aM359	249	2	aM409	308	5	jF463	362	2
aF360	265	2	aF410	313	5	jM464	361	4
aF361	266	5	aM411	314	5	aF465	360	2
aM362	274	5	aM412	311	2	aM466	364	2
aM363	257	5	aM413	312	2	jM467	366	2
aF364	255	5	aM414	315	2	jM468	365	2
aM365	259	5	aM415	310	2	jM469	363	2
aF366	264	5	aM416	306	5	aM470	367	7
aM367	261	5	aF417	307	2	jM471	368	2
aM368	258	5	aF418	309	5	jM472	369	2
aM369	262	5	aM419	316	2	jF473	374	2
aM370	271	2	aF420	317	2	jM474	375	2
am371	269	2	aM421	319	2	aF475	376	2
aF372	270	2	aM422	318	2	aF476	372	2
			aF423	320	2	jM477	377	2

Appendix E (continued)

Identification Number	Catalog Number	Vocal Group	Identification Number	Catalog Number	Vocal Group	Identification Number	Catalog Number	Vocal Group
aF478	370	2	aF531	435	2	jF587	499	4
jM479	373	2	aF532	433	2	jF588	497	4
jF480	378	2	aF533	437	2	aM589	502	2
jF481	382	2	aF534	434	2	jM590	503	2
jM482	379	2	aM535	429	2	jF591	506	2
jM483	384	2	aF536	438	2	jM592	504	2
jF484	381	2	aM537	439	2	aM593	501	2
jM485	380	2	aF538	441	2	aF594	498	2
aF486	383	2	aM539	440	2	aF595	507	4
jF487	385	2	aF540	442	2	jM596	505	2
aM488	391	2	aF541	443	2	jF597	508	4
jF489	387	2	aM542	444	2	aM598	509	2
jF490	388	2	aF543	445	2	aM599	511	2
jM491	390	5	aF544	446	2	jF600	510	2
aM492	386	2	aF545	449	2	aM601	512	4
jM493	389	2	aM546	447	2	jF602	514	7
aF494	392	5	aM547	448	2	jM603	518	3
jM495	393	2	aM548	457	2	jF604	517	3
jM496	394	5	aM549	454	2	jM605	519	3
aF497	401	7	aM550	455	2	jM606	520	4
jM498	396	7	aM551	456	2	jM607	516	4
jM499	400	5	aF552	450	2	jF608	515	4
jM500	402	5	aF555	453	2	jF609	521	4
jM501	398	5	aM556	458	2	aM610	522	2
aM502	403	2	aF557	459	2	aF611	523	2
aM503	397	2	jM558	460	4	aM612	524	2
jM504	399	2	aM559	462	4	aM613	525	5
aF505	404	2	aF560	461	4	aM614	530	2
aM506	405	2	aM561	463	5	aM615	527	3
aM507	406	2	aF562	464	4	aM616	528	4
jF508	407	2	aF563	465	4	jM617	529	4
aF509	408	2	aF564	466	4	jM618	531	4
jF510	409	2	aM565	467	4	jM619	532	4
aM514	418	2	aF566	468	2	jF620	533	3
aM515	422	2	aF567	475	4	aM621	534	3
aM516	417	2	aM568	471	4	aM622	535	2
aF517	423	2	aM569	483	5	aF623	536	2
jF518	421	2	aF570	476	2	jF624	538	4
aM519	420	2	aF571	478	5	aF625	537	4
jF520	416	2	aM572	479	4	aF626	543	5
jF521	415	2	aM573	477	5	jF627	541	7
aM522	419	2	jM577	484	5	jF628	539	7
aF523	425	2	aF578	485	5	jF629	544	7
aM524	426	2	aF579	486	4	aM630	540	2
aM525	432	2	aM580	487	2	aM631	542	2
aF526	427	2	aF581	488	2	aF632	547	3
aM527	430	2	aM582	489	2	jF633	545	3
aF528	431	2	aM583	494	3	aM634	546	2
aM529	428	2	aF584	495	3	jM635	553	1
aM530	436	2	aM585	496	3	aF636	548	1
			aM586	500	3			

Appendix E (continued)

Identification Number	Catalog Number	Vocal Group	Identification Number	Catalog Number	Vocal Group	Identification Number	Catalog Number	Vocal Group
jF637	549	3	aM680	606	2	aM722	649	2
jF638	550	4	aM681	613	2	aF723	658	2
jF639	555	3	aF682	609	2	aM724	657	2
aM640	551	3	aF683	607	2	aM725	655	2
jF641	556	3	aF684	610	2	aM726	652	2
jM642	554	3	aM685	611	2	aF727	653	2
aM643	557	2	aM686	617	2	aM728	654	2
aF644	552	2	aF687	618	2	aM729	656	2
jF645	561	7	aM688	622	2	aF730	662	2
jF646	562	7	aM689	612	2	aM731	659	2
jF647	563	7	aM690	614	2	aF732	660	2
jF648	564	7	aF691	615	2	aM733	661	2
aF649	560	2	aM692	616	2	aM734	663	2
aM650	559	2	aM693	623	2	aM735	664	2
aM651	565	4	aM694	619	2	aM736	665	2
aF652	566	4	aF695	620	2	jM737	666	2
aM653	569	2	aF696	621	2	aF738	667	2
jM564	570	2	aF697	624	2	aM739	668	2
aM655	571	2	aM698	628	4	aF740	669	2
aM656	572	2	aF699	627	4	aM741	670	2
aM657	575	2	aF700	629	3	aM742	673	2
aF658	576	2	aM701	634	7	aF744	678	6
aM659	577	2	aF702	638	2	aF745	705	6
aF660	578	2	aM703	635	2	aM746	679	6
aM661	579	2	aF704	637	2	aM747	682	6
aF662	580	2	aM705	630	5	aF748	681	6
aF663	583	5	aM706	636	7	aM749	692	6
aM664	594	5	aM707	631	5	aM750	683	6
aM665	581	5	aF708	632	5	aM751	684	6
aM666	586	5	aM709	633	5	aF752	685	6
aF667	587	5	aM710	640	5	aM753	686	6
aM668	590	5	aM711	639	7	aF754	687	6
aF669	582	5	aF712	641	5	aF755	688	6
aF670	584	5	aM713	642	5	aM756	689	6
aF672	588	5	aM714	644	2	aM757	690	6
aF673	593	5	aM715	643	2	aF758	697	2
aM674	592	5	aM716	648	2	aF759	700	2
aM675	589	5	aF717	647	5	aM760	701	2
aF676	595	5	aM718	645	5	jM761	702	2
aM677	591	5	aM719	646	7	jM762	699	2
aM678	601	2	aM720	651	2	jF763	698	2
aM679	608	2	jF721	650	2			

a. Numbers refer to identification numbers recorded on tapes and audiospectrograms. Coding as follows: a = adult, j = juvenile (age scores 0–6, Table 1); M = male, F = female, O = sex undetermined.
b. JGG catalog numbers on specimen tags; nc = not collected.
c. Numbers refer to vocal group designations given in the text.

Appendix F. Allozyme Frequencies for Polymorphic Loci in *L. leucoptera* and 7 Samples of *L. curvirostra*

Locus	Allele	Type 1	Type 2	Type 3	Type 4	Type 5	Type 6	Type 7	leucoptera
N:		28	308	60	74	61	15	15	16
ICD-1	167		0.002		0.004				
	148								0.031
	100	0.643	0.768	0.775	0.757	0.787	0.800	0.700	0.375
	88								0.094
	49	0.357	0.231	0.225	0.230	0.213	0.200	0.300	0.500
MPI	111			0.008					
	107		0.003		0.020		0.033	0.067	
	100	0.929	0.956	0.958	0.926	0.943	0.967	0.900	1.000
	92	0.071	0.041	0.025	0.054	0.057		0.033	
	82			0.008					
α-GPD	130	0.054	0.028	0.033	0.027	0.016		0.033	0.031
	100	0.929	0.946	0.958	0.946	0.975	1.000	0.900	0.906
	87		0.003					0.033	
	78		0.003						
	68	0.018	0.019	0.008	0.027	0.008		0.033	0.063
NP	140	0.036	0.006	0.025	0.007	0.008			0.031
	115	0.036	0.003	0.025	0.014	0.008			
	108	0.911	0.976	0.950	0.020	0.975	1.000	0.033	0.094
	100		0.015		0.953			0.967	0.875
	89								
	84	0.018							

Appendix F (continued)

Locus	Allele	Type 1	Type 2	Type 3	Type 4	Type 5	Type 6	Type 7	leucoptera
PGM	118								0.031
	113		0.005	0.008		0.008			
	100	1.000	0.003	0.983	0.993	0.984	1.000	1.000	0.969
	78		0.979	0.008	0.007				
	59		0.013	0.008		0.008			
GPI	165		0.010		0.034	0.008		0.067	
	158		0.008	0.050	0.014	0.025	0.067		
	152						0.933	0.933	1.000
	100	1.000	0.977	0.950	0.946	0.967			
	80		0.002						
	66		0.002						
	54		0.002		0.007				
CK-2	151	0.964	0.005		0.007	0.992	0.033		
	100	0.036	0.994	1.000	0.993	0.008	0.967	1.000	1.000
	59		0.002						
EST-4	170		0.990	1.000	0.993	1.000	0.967	1.000	0.094
	100	1.000	0.010		0.007		0.033		0.906
	39								
LA-1	130	0.018	0.021	0.017	0.014	0.025	0.067	0.033	
	124		0.003	0.008	0.007				
	120			0.008					
	115	0.089	0.141	0.067	0.095	0.123	0.133	0.100	0.063
	109		0.005	0.008	0.007	0.025			
	106		0.002		0.020				
	100	0.893	0.821	0.883	0.851	0.828	0.800	0.867	0.938
	91		0.006	0.008	0.007				

Appendix F (continued)

Locus	Allele	Type 1	Type 2	Type 3	Type 4	Type 5	Type 6	Type 7	leucoptera
LA-2	116		0.003		0.007				
	110	0.304	0.287	0.375	0.405	0.303	0.233	0.267	
	100	0.696	0.696	0.600	0.588	0.680	0.700	0.733	0.500
	93		0.002						
	91			0.008			0.033		
	87		0.005	0.017		0.008	0.033		0.438
	79		0.002						
	72		0.002						
	68		0.003			0.008			0.063
LGG	122	0.036	0.006	0.017	0.020	0.336	0.233	0.467	0.094
	111	0.554	0.369	0.450	0.432	0.656	0.767	0.533	0.719
	100	0.393	0.575	0.492	0.520	0.008			0.156
	88	0.018	0.049	0.042	0.027				0.031
	77		0.002						
PAP	108		0.003		0.007				
	105	1.000	0.997	0.000	0.993	1.000	1.000	1.000	1.000
GOT-1	209		0.003						0.031
	181		0.002						0.406
	100	0.982	0.990	0.992	1.000	1.000	1.000	1.000	0.469
	48								0.063
	19	0.018	0.005	0.008					0.031
GOT-2	0		0.003						
	-38		0.006	0.017			0.033		
	-100	1.000	0.990	0.983	1.000	1.000	0.967	1.000	1.000

Appendix F (continued)

Locus	Allele	Type 1	Type 2	Type 3	Type 4	Type 5	Type 6	Type 7	leucoptera
ACON-1	160		0.011		0.013	0.014		0.083	
	100	1.000	0.926	0.914	0.921	0.943	0.938	0.833	1.000
	53		0.063	0.086	0.066	0.043	0.063	0.083	
6-PGD	100	0.929	0.875	0.892	0.905	0.844	0.667	0.833	0.844
	87	0.071	0.123	0.108	0.095	0.156	0.333	0.167	
	81		0.002						0.156
SOD-1	179		0.006	0.008					0.031
	100	0.982	0.992	0.983	1.000	0.992	0.967	1.000	0.719
	81								0.063
	71			0.008					
	15	0.018	0.002			0.008	0.033		0.188
ACP	109	0.018	0.057	0.117	0.101	0.016		0.100	0.063
	100	0.982	0.942	0.883	0.899	0.984	1.000	0.900	0.938
	87		0.002						

Literature Cited

Adkisson, C. S.
    1981.   Geographic variation in vocalizations of North American pine grosbeaks. Condor 83:277-288.

Alatalo, R. V., L. Gustafsson, and A. Lundberg
    1984.   Why do young passerine birds have shorter wings than older birds? Ibis 126:410–415.

Alatalo, R. V., and A. Lundberg
    1986.   Heritability and selection on tarsus length in the pied flycatcher (*Ficedula hypoleuca*). Evolution 40:574–583.

American Ornithologists' Union
    1931.   Check-list of North American Birds, 4th ed. American Ornithologists' Union, Lancaster, Penn.
    1957.   Check-list of North American Birds, 5th ed. American Ornithologists' Union, Baltimore.

Avise, J. C., and C. F. Aquadro, and J. C. Patton
    1982.   Evolutionary genetics of birds. V. Genetic distances within Mimidae (mimic thrushes) and Vireonidae (vireos). Biochem. Genetics 20:95–104.

Avise, J. C., J. C. Patton, and C. F. Aquadro
    1980a.   Evolutionary genetics of birds. I. Relationships among North American thrushes and allies. Auk 97:135–147.
    1980b.   Evolutionary genetics of birds. Comparative molecular evolution of New World warblers and rodents. J. Heredity 712:303–310.

Avise, J. C., and R. M. Zink
   1988.   Molecular genetic divergence between avian sibling species: king and clapper rails, long-billed and short-billed dowitchers, boat-tailed and great-tailed grackles, and tufted and black-crested titmice. Auk 105:516–528.

Ayala, F. J.
   1986.   On the virtues and pitfalls of the molecular evolutionary clock. J. Heredity 77:226–235.

Bailey, A. M., R. J. Neidrach, and A. L. Bailey
   1953.   The red crossbills of Colorado. Museum Pictorial 9, Denver Museum of Natural History.

Baird, S. F., T. M. Brewer, and R. Ridgway
   1874.   A History of North American Birds, vol. 1.

Baker, A. J., M. D. Dennison, A. Lynch, and G. LeGrand
   1990.   Genetic divergence in peripherally isolated populations of chaffinches in the Atlantic islands. Evolution 44:981–999.

Baker, A. J., and A. Moeed
   1987.   Rapid genetic differentiation and founder effect in colonizing populations of common mynas (*Acridotheres tristis*). Evolution 41:525–538.

Baker, M. C., D. B. Thompson, G. L. Sherman, M. A. Cunningham, and D. F. Tomback
   1982.   Allozyme frequencies in a linear series of song dialect populations. Evolution 36:1020–1029.

Barrowclough, G. F., K. W. Corbin, and R. M. Zink
   1981.   Genetic differentiation in the Procellariiformes. Comp. Biochem. Physiol. 69B:629–632.

Barrowclough, G. F., N. K. Johnson, and R. M. Zink
   1985.   On the nature of genic variation in birds, pp. 135–154 *in* Current Ornithology, vol. 2. Plenum Press, New York.

Barton, N. H., and B. Charlesworth
   1984.   Genetic revolutions, founder effects, and speciation. Ann. Rev. Ecol. Syst. 15:133–164.

Baverstock, P. R., M. Adams, R. W. Polkinghorne, and M. Gelder
- 1982. The sex-linked enzyme in birds—Z-chromosome conservation but no dosage compensation. Nature 296:763–766.

Becker, P. H.
- 1982. The coding of species-specific characteristics of bird sounds, pp. 214–252 *in* D. E Kroodsma and E. H. Miller, eds., Acoustic Communication in Birds, vol. 1. Academic Press, New York.

Benkman, C. W.
- 1987. Food profitability and the foraging ecology of crossbills. Ecol. Monogr. 57:251–267.
- 1989. On the evolution and ecology of island populations of crossbills. Evolution 43:1324–1330.
- 1990. Intake rates and the timing of crossbill reproduction. Auk 107:376–386.

Bent, A. C.
- 1912. A new subspecies of crossbill from Newfoundland. Smithsonian Misc. Coll. 60:1–3.
- 1968. Red crossbill, pp. 497–526 *in* O. L. Austin, ed., Life Histories of North American Cardinals, Grosbeaks, Buntings, Towhees, Finches, Sparrows, and Allies. U. S. National Mus. Bull. 237.

Boag, P. T., and P. R. Grant
- 1978. Heritability of external morphology in Darwin's finches. Nature 274:793–794.
- 1981. Intense natural selection in a population of Darwin's finches (Geospizinae) in the Galápagos. Science 214:82–85.

Bock, C. E., and L. W. Lepthien
- 1976. Synchronous eruptions of boreal seed-eating birds. Am. Nat. 110:559–571.

Bonner, J. T.
- 1980. The Evolution of Culture in Animals. Princeton Univ. Press, Princeton.

Bowman, R. I.
- 1961. Morphological differentiation and adaptation in the Galápagos finches. Univ. Calif. Publ. Zool. 58:1–302.
- 1979. Adaptive morphology of song dialects in Darwin's finches. J. Ornithol. 120:353–389.

Braun, M. J., and M. B. Robbins
- 1986. Extensive protein similarity of the hybridizing chickadees *Parus atricapillus* and *P. carolinensis*. Auk 103:667–675.

Brehm, C. L.
- 1845. Drei neue deutsche Vogelgarten, nicht Subspecies, sondern species, und eine Beschriebung der bindigen Kreuzschnäbel. Isis von Oken 2:col. 243–269.
- 1846. Naturhistorische Bemerkungen über Nordamerika. Allgemeine Deutsche Naturhistorische Zeitung Dresden 1:530–536.
- 1853. Die Kreuzschnäbel. *Crucirostra,* Cuv. Naumannia 3:178–203.

Breninger, G. F.
- 1894. American and Mexican crossbills. Nidiologist 1:99–101.

British Ornithologists' Union
- 1934. List of British birds. Ibis 4:632–638.
- 1956. Report of the taxonomic committee. Ibis 98:157–168.

Brush, A. H., and D. M. Power
- 1976. House finch pigmentation: carotenoid metabolism and the effect of diet. Auk 93:725–739.

Bryant, E. H.
- 1986. On the use of logarithms to accommodate scale. Syst. Zool. 35:552–559.

Bush, G. L.
- 1975. Modes of animal speciation. Ann. Rev. Ecol. Syst. 6:339–364.

Carson, H. L., and A. R. Templeton
- 1984. Genetic revolutions in relation to speciation phenomena: the founding of new populations. Ann. Rev. Ecol. Syst. 15:97–131.

Cavalli-Sforza, L. L., M. W. Feldman, K. H. Chen, and S. M. Dornbusch
- 1982. Theory and observation in cultural transmission. Science 218:19–27.

Chapin, J. P.
- 1949. Pneumatization of the skull in birds. Ibis 91:691.

Coues, E.
- 1874. Birds of the Northwest. U. S. Government Printing Office, Washington, D. C.

Coutlee, E. L.
- 1971. Vocalizations in the genus *Spinus*. Animal Behav. 19:556–565.

Cracraft, J.
- 1983. Species concepts and speciation analysis. pp. 159–187 *in* R. F. Johnston, ed., Current Ornithology, vol. 1. Plenum Press, New York.
- 1989. Speciation and its ontology: the empirical consequences of alternative species concepts for understanding pattern and process of differentiation, pp. 28–59 *in* D. Otte and J. A. Endler, eds., Speciation and Its Consequences. Sinauer, Sunderland, Mass.

Davis, J.
- 1954. Seasonal changes in bill length of certain passerine birds. Condor 56:142–149.
- 1961. Some seasonal changes in morphology of the rufous-sided towhee. Condor 63:313–321.

Davis, M. B.
- 1983. Holocene vegetational history of the eastern United States, pp. 166–181 *in* H. E. Wright, Jr., ed., Late-Quaternary Environments of the United States, vol. 2, The Holocene. Univ. Minnesota Press, Minneapolis.

Delcourt, P. A., and H. R. Delcourt
- 1987. Late-Quaternary dynamics of temperate forests: applications of paleoecology to issues of global environmental change. Quaternary Science Reviews 6:129–146.

Dement'ev, G. P., Il'ichev, V. D., and E. V. Kurochkin
- 1965. (Functional analysis of some cases of polymorphism in birds and initial stages of speciation), pp. 141–155 *in* Vnutrividivaya Izmenichivost Pozvonochnykh Zhivotnykh I Mikroevol. Akad. Nauk, Sverdlovsk.

Diamond, J. M.
- 1990. Old dead rats are valuable. Nature 347:334–335.

Dickerman, R. W.
- 1986a. A review of the red crossbill in New York State, part 1, historical and nomenclatural background. Kingbird 36:73–78.
- 1986b. A review of the red crossbill in New York State, part 2, identification of specimens from New York. Kingbird 36:127–134.
- 1987. The "old northeastern" subspecies of red crossbill. Am. Birds 41:189–194.

Diehl, S. R., and G. L. Bush
  1989. The role of habitat preference in adaptation and speciation, pp. 345–365 *in* D. Otte and J. A. Endler, eds., Speciation and its Consequences. Sinauer Associates, Sunderland, Mass.

Dilger, W. C.
  1956. Adaptive modifications and ecological isolating mechanisms in the thrush genera *Catharus* and *Hylocichla.* Wilson Bull. 68:171–199.

Dwight, J., Jr.
  1900. The sequence of plumages and moults of the passerine birds of New York. N. Y. Acad. Sci. 13:73–360.

Echelle, A. A., and E. Kornfield
  1984. Evolution of fish species flocks. Univ. Maine Press, Orono.

Emlen, S. T.
  1972. An experimental analysis of the parameters of bird song eliciting species recognition. Behaviour 56:130–171.

Endler, J. A.
  1977. Geographic Variation, Speciation, and Clines. Princeton Univ. Press, Princeton, New Jersey.

Farris, J. S.
  1972. Estimating phylogenetic trees from distance matrices. Am. Nat. 106:645–668.
  1981. Distance data in phylogenetic analysis, pp. 1–23 *in* V. A. Funk and D. R. Brooks, eds., Advances in Cladistics, vol. 1. New York Botanical Garden, New York.

Felsenstein, J.
  1981. Skepticism toward Santa Rosalia, or why are there so few kinds of animals? Evolution 35:124–138.

Francis, C. M., and D. S. Wood
  1989. The effects of age and wear on wing length of warblers. J. Field Ornithol. 60:495–503.

Futuyma, D. J.
  1986. Evolutionary Biology, 2nd ed. Sinauer Associates, Sunderland, Mass.

Gill, F. B.
   1990.   Ornithology. Freeman, New York.

Gloger, C. L.
   1834.   Vollständiges Handbuch der Naturgeschichte der Vögel Europa's, mit besonderer Rücksicht auf Deutschland, vol. 1. August Schulz, Breslau.

Goldstein, R. B.
   1978.   Geographic variation in the "hoy" call of the bobwhite. Auk 95:85–94.

Gosler, A. G.
   1987.   Pattern and process in the bill morphology of the great tit *Parus major*. Ibis 129:451–476.

Grant, B. R., and P. R. Grant
   1979.   Darwin's finches: population variation, and sympatric speciation. Proc. Natl. Acad. Sci. USA 76:2359–2363.
   1989.   Evolutionary Dynamics of a Natural Population. Univ. Chicago Press, Chicago.

Grant, P. R.
   1981.   Patterns of growth in Darwin's finches. Proc. Roy. Soc. London B 212:403–432.
   1983.   Inheritance of size and shape in a population of Darwin's finches, *Geospiza conirostris*. Proc. Roy. Soc. London B 220:219–236.
   1986.   Ecology and Evolution of Darwin's Finches. Princeton Univ. Press, Princeton.

Grant, P. R., and B. R. Grant
   1989.   Sympatric speciation and Darwin's finches, pp. 433–457 *in* D. Otte and J. A. Endler, eds., Speciation and Its Consequences. Sinauer, Sunderland, Mass.

Greenwalt, C. H.
   1968.   Bird Song: Acoustics and Physiology. Smithsonian Inst. Press, Washington, D. C.

Grinnell, J.
   1909.   Birds and mammals of the 1907 Alexander expedition to southeastern Alaska. Univ. Calif. Publ. Zool. 5:171–264.

Griscom, L.
  1937. A monographic study of the red crossbill. Proc. Boston Soc. Nat. Hist. 41:77–210.

Groth, J. G.
  1988. Resolution of cryptic species in Appalachian red crossbills. Condor 90:745–760.
  1990. Cryptic species of nomadic birds in the red crossbill (*Loxia curvirostra*) complex of North America. Dissertation, Univ. California, Berkeley.
  1992. White-winged crossbill breeding in southern Colorado, with notes on juveniles' calls. Western Birds 23:35–37.
  1993. Call matching and positive assortative mating in red crossbills. Auk (in press).

Gutiérrez, R. J., R. M. Zink, and S. Y. Yang
  1983. Genic variation, systematic, and biogeographic relationships of some galliform birds. Auk 100:33–47.

Harris, H., and D. A. Hopkinson
  1978. Handbook of Enzyme Electrophoresis in Human Genetics. North Holland Publ. Co., Amsterdam.

Hartert, E.
  1904. Die Vögel der Paläarktischen Fauna, vol. 1. Friedlander, Berlin.

Hartert, E. and F. Steinbacher
  1932. Die Vögel der Paläarktischen Fauna. Friedländer, Berlin.

Hill, G. E.
  1992. Proximate basis of variation in carotenoid pigmentation in male house finches. Auk 109:1–12.

Hinde, R. A.
  1955. A comparative study of the courtship of certain finches. Ibis 97:706–745.

Horn, H. S., and R. M. May
  1977. Limits to similarity among coexisting competitors. Nature 270:660–661.

Howell, T. R.
  1972. Birds of the lowland pine savanna of northeastern Nicaragua. Condor 74:316–340.

Hutchinson, G. E.
- 1959. Homage to Santa Rosalia, or why are there so many kinds of animals? Am. Nat. 93:145–159.

Huxley, J. S.
- 1955. Morphism in birds. Proc. XI Intl. Ornithol. Congr.:309–328.

Immelmann, K.
- 1975. Ecological significance of imprinting and early learning. Ann. Rev. Ecol. Syst. 6:15–37.

Ince, S. A., P. J. B. Slater, and C. Weisman
- 1980. Changes with time in the songs of populations of chaffinches. Condor 82:285–290.

International Trust for Zoological Nomenclature
- 1985. International Code of Zoological Nomenclature, 3rd ed. International Trust for Zoological Nomenclature, London.

James, F. C.
- 1982. The ecological morphology of birds: a review. Ann. Zool. Fennici 19:265–275.
- 1983. Environmental component of morphological variation in birds. Science 221:184–186.

James, F. C., and C. E. McCulloch
- 1990. Multivariate analysis in ecology and systematics: panacea or Pandora's box? Ann. Rev. Syst. Ecol. 21:129–166.

Jenni, D. A., R. D. Gambs, and J. Betts
- 1974. Acoustic behavior of the northern jacana. Living Bird 13:193–210.

Johnson, N. K.
- 1963. Biosystematics of sibling species of flycatchers in the *Empidonax hammondii–oberholseri–wrightii* complex. Univ. Calif. Publ. Zool. 66:79–238.
- 1980. Character variation and evolution of sibling species in the *Empidonax difficilis–flavescens* complex (Aves: Tyrannidae). Univ. Calif. Publ. Zool. 112:1–151.

Johnson, N. K., and R. E. Jones
   1990.   Geographic differentiation and distribution of the Peruvian screech-owl. Wilson Bull. 102:199–212.

Johnson, N. K., and J. A. Marten
   1988.   Evolutionary genetics of flycatchers. II. Differentiation in the *Empidonax difficilis* complex. Condor 105:177–191.

Johnson, N. K., J. A. Marten, and C. J. Ralph
   1989.   Genetic evidence for the origin and relationships of the Hawaiian honeycreepers (Aves: Fringillidae). Condor 91:379–396.

Johnson, N. K., and R. M. Zink
   1983.   Speciation in sapsuckers (*Sphyrapicus*): I. Genetic differentiation. Auk 100:871–884.

Johnson, R. E.
   1977.   Seasonal variation in the genus *Leucosticte* in North America. Condor 79:76–86.

Jollie, M.
   1953.   Plumages, molts, and racial status of red crossbills in northern Idaho. Condor 55:193–197.

Kemper, T.
   1959.   Notes on the breeding cycle of the red crossbill (*Loxia curvirostra*) in Montana. Auk 76:181–189.

Kirikov, S. V.
   1940.   On the connection between the red crossbills and the coniferous trees. Bull. Acad. Sci. USSR (Biol.) 1940:359–376.

Knox, A. G.
   1975.   Crossbill taxonomy, pp. 191–201 *in* D. Nethersole-Thompson, Pine Crossbills. T. & A. D. Poyser, Berkhamsted.
   1976.   The taxonomic status of the Scottish crossbill, *Loxia* sp. Bull. Brit. Ornithol. Club 96:15–19.
   1983.   Handedness in the crossbills *Loxia* and the akepa *Loxops coccinea*. Bull. Brit. Ornithol. Club 103:114–118.
   1990.   The sympatric breeding of common and Scottish crossbills *Loxia curvirostra* and *L. scotica*, and the evolution of crossbills. Ibis 132:454–466.

Kroodsma, D. E.
    1981. Geographical variation and functions of song types in warblers (Parulidae). Auk 98:743–751.

Kroodsma, D. E., and J. R. Baylis
    1982. A world survey of evidence for vocal learning in birds, pp. 311–337 *in* D. E. Kroodsma and E. H. Miller, eds., Acoustic Communication in Birds, vol. 2. Academic Press, New York.

Kruskal, J. G.
    1964. Multidimensional scaling by optimizing goodness of fit to a nonmetric hypothesis. Psychometrika 29:1–27.

Lack, D.
    1944a. Ecological aspects of species formation in passerine birds. Ibis 86:260–286.
    1944b. Correlation between beak and food in the crossbill *Loxia curvirostra* Linneaus. Ibis 86:552–553.
    1971. Ecological Isolation in Birds. Blackwell Sci. Publ., Oxford.

Lanyon, W. E.
    1978. Revision of the *Myiarchus* flycatchers of South America. Bull. Am. Mus. Nat. Hist. 161:427–628.

Larson, A., E. M. Prager, and A. C. Wilson
    1984. Chromosomal evolution, speciation and morphological change in vertebrates: the role of social behavior. Chromosomes Today 8:215–228.

Lauder, G. V.
    1981. Form and function: structural analysis in evolutionary morphology. Paleobiology 7:430–442.

Lawrence, L. deK.
    1949. The red crossbill at Pimisi Bay, Ontario. Canadian Field-Nat. 63:147–160.

Linnaeus, C.
    1758. Systema Naturae, 10th ed. Laurentii Salvii, Holmaie.

Löhrl, H.
    1963. The use of bird calls in clarifying taxonomic relationship. Proc. XIII Intl. Ornithol. Congr. 544–552.

Løvtrup, S., F. Rahemtulla, and N.-G. Höglund
    1974. Fisher's axiom and the body size of animals. Zool. Scripta 3:53–58.

MacArthur, R. H., and R. Levins
    1967. The limiting similarity, convergence, and divergence of coexisting species. Am. Nat. 101:377–386.

MacNally, R. C., D. M. Weary, R. E. Lemon, and L. Lefebvre
    1986. Species recognition by song in the veery (*Catharus fuscescens*: Aves). Ethology 71:125–139.

Maier, V.
    1982. Acoustic communication in the guinea fowl (*Numidea meleagris*): structure and use of vocalizations and the principles of message coding. Z. Tierpsychol. 59:29–83.

Mantel, N.
    1967. The detection of disease clustering and a general regression approach. Cancer Res. 27:209–220.

Marler, P.
    1957. Specific distinctiveness in the communication signals of birds. Behaviour 11:13–39.

Marler, P., and P. C. Mundinger
    1975. Vocalizations, social organization, and breeding biology of the twite *Acanthus flavirostris*. Ibis 117:1–17.

Marten, J. A., and N. K. Johnson
    1986. Genetic relationships of North American cardueline finches. Condor 88:409–420.

Massa, B.
    1987. Variations in Mediterranean crossbills (*Loxia curvirostra*). Bull. Brit. Ornithol. Club 107:118–129.

Maynard Smith, J.
    1966. Sympatric speciation. Am. Nat. 100:637–650.

Mayr, E.
    1942. Systematics and the Origin of Species. Columbia Univ. Press, New York.
    1963. Animal Species and Evolution. Harvard. Univ. Press, Cambridge, Mass.

Mayr, E., and L. L. Short
 1970. Species taxa of North American birds. Publ. Nuttall Ornithol. Club 9:1–127.

Mayr, E., E. G. Linsley, and R. L. Usinger
 1953. Methods and Principles of Systematic Zoology. McGraw-Hill, New York.

McCabe, T. T., and E. B. McCabe
 1933. Notes on the anatomy and breeding habits of crossbills. Condor 35:136–147.

Meacham, C. A., and T. Duncan
 1990. MorphoSys Version 1.26. Regents of the University of California, Berkeley.

Meinertzhagen, R.
 1934. (Note on the parrot crossbill [*Loxia pytyopsittacus*]). Bull. Brit. Ornithol. Club 55:56–58.

Meinertzhagen, R., and K. Williamson
 1953. Check-list of the birds of Great Britain and Ireland. Ibis 95:365–369.

Messineo, D. J.
 1985. The 1985 nesting of pine siskin, red crossbill, and white-winged crossbill in Chenango County, N. Y. Kingbird 35:233–237.

Miller, A. H.
 1946. A method of determining the age of live passerine birds. Bird-Banding 17:33–35.

Miller, D. B.
 1977. Two-voice phenomenon in birds: further evidence. Auk 94:567–572.

Miller, E. H.
 1979. An approach to the analysis of graded vocalizations of birds. Behav. Neur. Biol. 27:25–38.
 1982. Character and variance shift in acoustic signals of birds, pp. 253–295 *in* D. E. Kroodsma and E. H. Miller, eds., Acoustic Communication in Birds, vol. 1. Academic Press, New York.
 1986. Components of variation in the nuptial calls of the least sandpiper (*Calidris minutilla;* Aves, Scolopacidae). Syst Zool. 35:400–413.

Monson, G., and A. R. Phillips
    1981. The races of red crossbill, *Loxia curvirostra,* in Arizona, pp. 223–230 *in* G. Monson and A. R. Phillips, Checklist of Birds of Arizona. Univ. Arizona Press, Tucson.

Morton, E. S.
    1975. Ecological sources of selection on avian vocalizations. Am. Nat. 109:17–34.
    1982. Grading, discreteness, redundancy, and motivation-structural rules, pp. 183–212 *in* D. E. Kroodsma and E. H. Miller, eds., Acoustic Communication in Birds, vol. 1. Academic Press, New York.

Mundinger, P. C.
    1970. Vocal imitation and individual recognition of finch calls. Science 168:480–482.
    1979. Call-learning in the Carduelinae: ethological and systematic implications. Syst. Zool. 28:270–283.
    1980. Animal cultures and a general theory of cultural evolution. Ethol. Sociobiol. 2:183–223.

Nei, M.
    1977. F-statistics and analysis of gene diversity in subdivided populations. Ann. Human Genet. 41:225–233.
    1978. Estimation of average heterozygosity and genetic distance from a small number of individuals. Genetics 89:583–590.

Nethersole-Thompson, D.
    1975. Pine Crossbills. T. & A. D. Poyser, Berkhamsted.

Newton, I.
    1970. Irruptions of crossbills in Europe, pp. 337–357 *in* A. Watson, ed., Animal Populations in Relation to their Food Resources. Blackwell Scientific, Oxford.
    1973. Finches. Taplinger, London.

Nicolai, J.
    1964. Der Brutparasitismus der Viduinae als ethologisches Problem. Prägungsphänomene als Faktoren der Rassen- und Artbildung. Z. Tierpsychol. 21:129–204.

Nuechterlein, G. L.
- 1981. Courtship behavior and reproductive isolation between western grebe color morphs. Auk 98:335–349.

Pääbo, S.
- 1989. Ancient DNA: extraction, characterization, molecular cloning, and enzymatic amplification. Proc. Natl. Acad. Sci. 86:1939–1943.

Packard, G. C.
- 1967. Seasonal changes in bill length of house sparrows. Wilson Bull. 79:345–346.

Paterson, H. E. H.
- 1985. The recognition concept of species, pp 21–29 in E. S. Vrba, ed., Species and Speciation. Transvall Mus. Monogr. 4. Transvall Museum, Pretoria.

Payne, R. B.
- 1973. Behavior, mimetic songs and song dialects, and relationships of the parasitic indigobirds (*Vidua*) of Africa. Ornithol. Monogr. 11.
- 1982. Species limits in the indigobirds (Ploceidae: *Vidua*) of West Africa: mouth mimicry, song mimicry, and description of new species. Misc. Publ. Mus. Zool. Univ. Michigan 162.
- 1986. Bird songs and avian systematics, pp. 87–126 in R. F. Johnston, ed., Current Ornithology, vol. 3. Plenum Press, New York.
- 1987. Populations and type specimens of a nomadic bird: comments on the North American crossbills *Loxia pusilla* Gloger 1834 and *Crucirostra minor* Brehm 1845. Occ. Pap. Mus. Zool. Univ. Mich., no. 714.

Payne, R. B., W. L. Thompson, K. L. Fiala, and L. L. Sweany
- 1981. Local song traditions in indigo buntings: cultural transmission of behavior patterns across generations. Behavior 77:199–221.

Peterson, J. M. C.
- 1985. Region 7—Adirondack–Champlain. Kingbird 35:139–142.

Phillips, A. "A." (=R.)
- 1975. The incredible American red crossbill *Loxia curvirostra* Carduelinae. Emu 74:282.

Phillips, A. R.
- 1977. Sex and age determination of red crossbills. Bird-Banding 48:110–117.

Phillips, A. R., J. Marshall, and G. Monson
    1964. The Birds of Arizona. Univ. Arizona Press, Tucson.

Pitocchelli, J.
    1990. Plumage, morphometric, and song variation in mourning (*Oporornis philadelphia*) and MacGillivray's (*O. tolmiei*) warblers. Auk 107:161–171.

Pratt, H. D.
    1989. Species limits in akepas (Drepanidinae: *Loxops*). Condor 91:933–940.

Price, T. D.
    1984. The evolution of sexual size dimorphism in Darwin's finches. Am. Nat. 123:500–518.

Pulliam, H. R.
    1985. Foraging efficiency, resource partitioning, and the coexistence of sparrow species. Ecology 66:1829–1836.

Raikow, R. J.
    1986. Why are there so many kinds of passerine birds? Syst. Zool. 35:254–255.

Ratcliffe, L. M., and P. R. Grant
    1985. Species recognition in Darwin's finches (*Geospiza*, Gould). I. Discrimination by morphological cues. Anim. Behav. 31:1139–1153.

Ratti, J. T.
    1979. Reproductive separation and isolating mechanisms between sympatric dark- and light-phase western grebes. Auk 96:573–586.

Reinikainen, A.
    1937. The irregular migration of the crossbill (*L. c. curvirostra*) and their relation to the cone crop of conifers. Ornis Fennica 14:55–64.

Remsen, J. V., Jr., K. Garrett, and R. A. Erikson
    1982. Vocal copying in Lawrence's and lesser goldfinches. Western Birds 13:29–33.

Rice, R. W.
    1984. Disruptive selection on habitat preference and the evolution of reproductive isolation: a simulation study. Evolution 38:1251–1260.

Richards, L. P., and W. J. Bock
  1973. Functional anatomy and the adaptive evolution of the feeding apparatus in the Hawaiian honeycreeper genus *Loxops* (Drepanididae). Ornithol. Monogr. 15.

Richardson, B. J., P. R. Baverstock, and M. Adams
  1986. Allozyme Electrophoresis. Academic Press, Sydney, Australia.

Ricklefs, R. E.
  1968. Patterns of growth in birds. Ibis 110:419–451.

Ridgway, R.
  1885a. A review of the American crossbills (*Loxia*) of the *L. curvirostra* type. Proc. Biol. Soc. Washington 2:101–107.
  1885b. Some emended names of North American birds. Proc. U. S. Natl. Mus., pp. 354–356.
  1912. Color standards and color nomenclature. R. Ridgway, Washington, D. C.

Robins, J. D., and G. D. Schnell
  1971. Skeletal analysis of the *Ammodramus–Ammospiza* grassland sparrow complex: a numerical taxonomic study. Auk 88:567–590.

Rockwell, R. F., and G. F. Barrowclough
  1986. Gene flow and the genetic structure of populations, pp. 223–255 *in* F. Cooke and P. A. Buckley, eds., Avian Genetics. Academic Press, New York.

Rogers, J. S.
  1972. Measures of genetic similarity and genetic distance. Univ. Texas Publ. 7213:145–153.
  1986. Deriving phylogenetic trees from allelic frequencies: a comparison of nine genetic distances. Syst. Zool. 35:297–310.

Rohlf, F. J., J. Kishpaugh, and D. Kirk
  1982. Numerical Taxonomy System of Multivariate Statistical Programs. State Univ. New York, Stony Brook.

Rowley, I.
  1967. A fourth species of Australian corvid. Emu 66:191–210.

Samson, F. B.
  1978. Vocalizations of Cassin's finch in northern Utah. Condor 80:203–210.

SAS Institute
  1985. SAS User's Guide: Statistics. SAS Institute, Inc., Cary, North Carolina.

Schluter, D., and P. R. Grant
  1984. Determinants of morphological patterns in communities of Darwin's finches. Am. Nat. 123:175–196.

Schnell, G. D., D. J. Watt, and M. E. Douglas
  1985. Statistical comparison of proximity matrices: applications in animal behavior. Anim. Behav. 33:239–253.

Schoener, T. W.
  1965. The evolution of bill size differences among sympatric congeneric species of birds. Evolution 19:180–213.

Schueler, F. W., and J. D. Rising
  1976. Phenetic evidence for natural hybridization. Syst. Zool. 25:283–289.

Selander, R. K.
  1958. Age determination and molt in the boat-tailed grackle. Condor 60:255–376.
  1964. Speciation in wrens of the genus *Campylorhynchus*. Univ. Calif. Publ. Zool. 74.

Selander, R. K., and R. F. Johnston
  1967. Evolution of the house sparrow. I. Intrapopulation variation in North America. Condor 69:217–258.

Selander, R. K., M. H. Smith, S. Y. Yang, W. E. Johnson, and J. B. Gentry
  1971. Biochemical polymorphism and systematics in the genus *Peromyscus*. I. Variation in the old-field mouse (*Peromyscus polionotus*). Univ. Texas Publ. 7103:49–90.

Shiovitz, K., and R. E. Lemon
  1980. Species identification by songs in the indigo bunting, *Passerina cyanea*, and its relationship to organization of avian acoustical behavior. Behaviour 55:128–179.

Slatkin, M.
  1981. Estimating levels of gene flow in natural populations. Genetics 99:323–335.
  1985a. Rare alleles as indicators of gene flow. Evolution 39:53–64.
  1985b. Gene flow in natural populations. Ann. Rev. Ecol. Syst. 16:393–430.

Slatkin, M., and N. H. Barton
- 1989. A comparison of three indirect methods for estimating average levels of gene flow. Evolution 43:1349–1368.

Smith, J. N. M., and R. Zach
- 1979. Heritability of some morphological characters in a song sparrow population. Evolution 33:460–467.

Smith, T. B.
- 1987. Bill size polymorphism and intraspecific niche utilization in an African finch. Nature 329:717–719.
- 1990a. Comparative breeding biology of the two bill morphs of the black-bellied seedcracker. Auk 107:153–160.
- 1990b. Natural selection on bill characters in the two bill morphs of the African finch *Pyrenestes ostrinus*. Evolution 44:832–842.

Sneath, P. H. A., and R. R. Sokal
- 1973. Numerical Taxonomy. W. H. Freeman, San Francisco.

Sparling, D. W., and J. D. Williams
- 1978. Multivariate analysis of avian vocalizations. J. Theor. Biol. 74:83–107.

Stein, R. C.
- 1958. The behavioral, ecological, and morphological characteristics of two populations of the alder flycatcher, *Empidonax traillii* (Audubon). New York State Mus. Bull. 37:1–63.
- 1963. Isolating mechanisms between populations of Traill's flycatchers. Proc. Am. Philos. Soc. 107:21–50.

Stepanyan, L. S.
- 1979. An American race of the red crossbill (*Loxia curvirostra sitkensis*) as a casual visitor to the Kamchatka. Ornitologiya 14:198.

Stresemann, E.
- 1922. Der Name des Kreuzschnabels der amerikanischen Oststaaten. Ornithol. Monats. 30:41–42.

Strickland, A.
- 1851. (Note on Mexican crossbills). Jardine's Contrib. Ornithol. 1:43.

Swofford, D. L.
  1981. On the utility of the distance Wagner procedure, pp. 25–43 *in* V. A. Funk and D. R. Brooks, eds., Advances in Cladistics. New York Bot. Gardens, New York.

Swofford, D. L., and R. K. Selander
  1981. BIOSYS-1: A FORTRAN program for comprehensive analysis of electrophoretic data in population genetics and systematics. J. Hered. 72:281–283.

Tauber, C. A., and M. J. Tauber
  1977. A genetic model for sympatric speciation through habitat diversification and seasonal isolation. Nature 268:702–705.
  1989. Sympatric speciation in insects:perception and perspective, pp. 307–344 *in* D. Otte and J. A. Endler, eds., Speciation and Its Consequences. Sinauer Associates, Sunderland, Massachusetts.

Theilcke, G.
  1973. On the origin of divergence of learned signals (songs) in isolated populations. Ibis 115:511–516.

Thomas, W. K., S. Pääbo, F. X. Villablanca, and A. C. Wilson
  1990. Spatial and temporal continuity of kangaroo rat populations shown by sequencing mitochondrial DNA. J. Molecular Evolution 31:101–112.

Ticehurst, C. B.
  1915. On the plumages of the male crossbill (*Loxia curvirostra*). Ibis 3:662–669.

Tordoff, H. B.
  1952. Notes on plumages, molts, and age variation in the red crossbill. Condor 55:200–203.
  1954a. Social organization and behavior in a flock of captive, nonbreeding red crossbills. Condor 56:416–422.
  1954b. Further notes on plumages and molts of red crossbills. Condor 56:108–109.

Tordoff, H. B., and W. R. Dawson
  1965. The influence of daylength on reproductive timing in the red crossbill. Condor 67:416–422.

Traylor, M. A.
  1979. Two sibling species of *Tyrannus* (Tyrannidae). Auk 96:221–233.

Troy, D. M.
    1985. A phenetic analysis of the redpolls *Carduelis flammea flammea* and *C. hornemanni exilpes*. Auk 102:82–95.

van Noordwijk, A. J., J. H. van Balen, and W. Scharloo
    1980. Heritability of ecologically important traits in the great tit. Ardea 68:193–203.

van Rossem, A. J.
    1934. Notes on some types of North American birds. Trans. San Diego Soc. Nat. Hist. 7:347–362.

Vaurie, C.
    1959. Birds of the Palearctic Fauna. Witherby, London.

Vawter, L., and W. M. Brown
    1986. Nuclear and mitochondrial DNA comparisons reveal extreme rate variation in the molecular clock. Science 234:194–196.

Voous, K. H.
    1960. Atlas of European Birds. Nelson, London.
    1978. The Scottish crossbill: *Loxia scotica*. Brit. Birds. 71:3–10.

Wahlund, S.
    1928. Zuzammensetzung van Populationen und Korrelationserscheinungen vom Standpunkt der Veterbungslehre aus betrachtet. Hereditas 11:65–106.

Wallace, B.
    1981. Basic Population Genetics. Columbia Univ. Press, New York.

Wells, D. R.
    1982. Biological species limits in the *Cettia fortipes* complex. Bull. Brit. Ornithol. Club 102:62–63.

Wells, P. V.
    1970. Postglacial vegetational history of the Great Plains. Science 167:1574–1582.

West, G. C.
    1974. Abnormal bill of a white-winged crossbill. Auk 91:624–626.

Wheelwright, H. W.
- 1862. On the change of plumage in the crossbills and pine grosbeak. Zoologist:8001–8002.
- 1871. A Spring and Summer in Lapland, 2nd ed. Groombridge and Sons, London.

White, G.
- 1789. The Natural History of Selborne. Reprinted by G. Routledge & Sons, London.

Wilson, A.
- 1811. American Ornithology, or the Natural History of the Birds of the United States, vol. 4. Bradford and Insekeep, Philadelphia.

Witherby, H. F., F. C. R. Jourdain, N. F. Ticehurst, and B. W. Tucker
- 1938. The Handbook of British Birds, vol. 1. Witherby, London.

Wood, M.
- 1969. A Bird-Bander's Guide to the Determination of Age and Sex in Selected Species. Pennsylvania State Univ., University Park, Penn.

Workman, P. L., and J. D. Niswander
- 1970. Population studies on southwestern Indian tribes. II. Local genetic differentiation in the Papago. Am. J. Human Genet. 22:22–49.

Wright, S.
- 1951. The genetical structure of populations. Ann. Eugen. 15:323–354.
- 1965. The interpretation of population structure by F-statistics, with special regard to systems of mating. Evolution 19:395–420.
- 1978. Evolution and Genetics of Populations, vol. 4, Variability Within and Among Natural Populations. Univ. Chicago Press, Chicago.

Wyles, J. S., J. G. Kunkel, and A. C. Wilson
- 1983. Birds, behavior, and anatomical evolution. Proc. Natl. Acad. Sci. 80:4394–4397.

Yang, S. Y., and J. L. Patton
- 1981. Genetic variability and differentiation in the Galápagos finches. Auk 98:230–242.

Yunick, R. P.
    1977. Timing of completion of skull pneumatization in the pine siskin. Bird-Banding 48:67–71.

Zink, R. M.
    1988. Evolution of brown towhees: allozymes, morphometrics and species limits. Condor 90:72–82.
    1991. The geography of mitochondrial DNA variation in two sympatric sparrows. Evolution 45:329–339.

Zink, R. M., and N. K. Johnson
    1984. Evolutionary genetics of flycatchers: I. Sibling species in the genera *Empidonax* and *Contopus*. Syst. Zool. 33:205–216.

Zink, R. M., D. F. Lott, and D. W. Anderson
    1987. Genetic variation, population structure, and evolution of California quail. Condor 89:395–405.

Zink, R. M., and J. V. Remsen, Jr.
    1986. Evolutionary processes and patterns of geographic variation in birds, pp. 1–69 *in* R. F. Johnston, ed., Current Ornithology, vol. 4. Plenum Press, New York.

WITHDRAWN